THE ORIGIN A

A COMPLEX UN
DIRECT OBSERVATIONS

Bridge for Human Progress

An Advanced Understanding of Mind
More Than a Synthesis of the Sciences
Proves That There Is a Purpose and Design
to the Universe

By

LEONARD WILSON, JR. M. ED.
PSYCHOLOGIST/GENERAL SCIENTIST

DORRANCE
PUBLISHING CO
EST. 1920
PITTSBURGH, PENNSYLVANIA 15238

Dorrance Publishing Co
585 Alpha Drive
Suite 103
Pittsburgh, PA 15238
Visit our website at *www.dorrancebookstore.com*

ISBN: 978-1-6376-4320-4
eISBN: 978-1-6376-4634-2

THE ORIGIN AND NATURE OF MIND

A COMPLEX UNDERSTANDING FROM DIRECT OBSERVATIONS

Bridge for Human Progress

**An Advanced Understanding of Mind
More Than a Synthesis of the Sciences
Proves That There Is a Purpose and Design
to the Universe**

By

LEONARD WILSON, JR. M. ED.
PSYCHOLOGIST/GENERAL SCIENTIST

CONTENTS

INTRODUCTION

Today's world, including the United States, is plagued by an assortment of debilitating mental disorders. In fact, one specific condition known as autism is constantly increasing. Prior to November 2015, the CDC announced that one in 68 children in the United States had a developmental disability in 2006-2008.The total cost per year for children with ASD (autism spectrum disorder) in the United States was estimated to be between $11.5 billion-$60.9 billion (2011 US dollars). As of November 2015, the CDC announced that one in 45 children in the US has ASD using new and improved diagnostic methods.The Population Reference Bureau reported that the United States has the highest incarceration rate in the world.The US rate is 500 prisoners per 100,000 residents. My experience from working in the penal system for 15 years is that a large number of inmates are undiagnosed learning disabled, mentally ill, and ASD victims. Our school system is not equipped to deal with this. In addition, there is an ever-increasing number of homeless people in this nation, and many of them are mentally ill and suffer addictions. Our modern institutions, armed with an ever increasing body of knowledge, seems helpless! Adding to the problem is the fact that the senior citizen population has also grown at a rapid pace.With this comes more debilitating mental problems, such as dementia, Alzheimer's and Parkinson's disease. Medical care in general is going to have to drastically improve, as illness is a double edge sword, costing billions in lost productivity and billions for healthcare.

These factors will interface over time and will manifest as a national disaster. As the disabled segment of the population becomes larger, national security and the economy will suffer.The dwindling work force won't be able to support the nation. Every segment of society will be overburdened, especially the family unit which is already crumbling.

It is my hope to put an end to the needless suffering of millions of individuals and their families and the great disaster facing this nation.This book is an effort to enhance the understanding of the mind and establish a foundation for a new approach to doing mind research. I have used an approach that involves a synthesis of the natural sciences rather than looking at the mind from one perspective. I have found that looking at the mind from multiple perspectives enhances understanding due to the additional information brought to bear on the problem. It is my belief that modern science has developed enough information to solve many of our most serious problems and that a new problem-solving paradigm using current knowledge will suffice.

In this approach, I am first of all going to reconnect with physics in more than one way. Leading scientists such as Stephen Hawking and Leonard Mlodinow (2010) believe that it is necessary to use models as frame works for understanding physical realities (p.7). The model method is how my work is divergent from traditional psychology, psychiatry, and neuroscience.

Finally, even though mind is a serious problem facing the United States, it is my hope that minorities such as black Americans will experience immediate relief. This book is intended to help all Americans. Those with the least resources are suffering the most and are in the most, dire need of help. The most important resource is information. Blacks with unique melanin physiology will suffer most when confronted with the conditions presented to them just by living in the United States: geography, weather, transportation technology, building design, exercise science, food science, and the narrow scope of medical science.

I would also like to point out that this book isn't an attack on modern science or Christianity. If it wasn't for math and science, I couldn't understand the Bible, and if not for Christianity, I wouldn't have a Bible. This isn't an attack on any group of people, but an attempt to save humanity from extinction. Most of all this is not a rebuke, but rather a revealing and a call to action. This is my contribution! Most importantly, this is an effort for support to extend human rights to the unborn. A plea for the implementation of informed, comprehensive measures, that ensure that all pregnancies in America will produce healthy babies with the potential of becoming productive adults with mental health; a fulfillment of my obedience to God and responsibility to man. Modern biology has made great strides, and this is a natural extension and keep in mind this is an introductory work. There is much more to be revealed!

Chapter 1

THE BRAIN FRAME OF REFERENCE

The modem era of research has focused on developing sophisticated technology to help observe and explain complex biological phenomenon. Even so, some questions remain unanswered. One major question is, **what is mind**? Mind scientists have almost limited their study of mind to the organ known as the brain. Scientists who study the brain do so by observing the physical and chemical activities within the brain itself and human behaviors that seem to be related. Will this approach lead to the answer? Is this question so complex that science cannot answer it? Can it be approached from another perspective? What is the proper approach?

Myers and Jeeves quoted Polkinghome as follows: "Reality is multi-layered unity. I can perceive another person as aggregation of atoms, an open biochemical system in interaction with the environment, a specimen of homo sapiens" (Myers and Jeeves, (1987, p.5).

Myers and Jeeves (1987) made the following statement:

> *What is true of psychology is also true of the other academic disciplines, each of which provides a perspective from which we can study nature and our place in it. These range from the scientific fields that study the most elementary building blocks of nature up to philosophy and theology which address some of life's global questions.* (p.8)

Mind is our global question, and we are in search of a perspective that will allow us to describe it.

The Most Important Subject

In addition to the enormous financial burden on this nation's economy as a result of the large number of people suffering from mental illness, learning disabilities, and behavioral problems, there are personal and social costs that are devastating. These afflictions overwhelm individuals, families, communities, schools, institutions, and law enforcement. Individuals afflicted with any form of mental illness aren't going to have a normal, productive life, and their families are going to have a life of hell along with them. The security of the entire nation is at risk. Dedicated men and women of science are putting forth herculean effort to solve this problem without success. However, I think it can safely be said that sufficient scientific knowledge exists for us to understand mind. We need new guiding principles and more sophisticated paradigms that result in research methodologies that aid our search and guarantee our success.

Let us look at this important quote from J.P. Guildford as sited by Jackson (1993): "The mind," says professor J.P. Guildford, "if it has any scientific meaning at all, is the functioning of the brain. But since the brain cannot function alone, but only in connection with nerves, receptors and effectors, mind is better described as the functioning of the entire nervous system" (p.48)

This is a very important insight on the way to understanding mind: a fundamental building block. It is the beginning of the formation of an entire paradigm. We must have a correct understanding of mind in order to successfully solve the problems associated with it.

This approach immediately allows us to look at mind without reference to biochemistry which has been the thrust of most brain research. Biochemistry is important but, it may not be a fundamental principle. Thus, concentrating on biochemistry may be a hindrance to understanding mind.

So far, we have only touched upon the adverse side of mind. However, there are great advantages to a functional mind. Jackson implicated that it was the evolution of the human mind that gave man his survival skills to cope with the harsh conditions that have confronted all biological life forms on planet earth throughout history. Fischbach (1993) put it this way, "our survival and probably the survival of this planet depend on a more complete understanding of the human mind" (p.1). A more complete understanding is another key insight and implies that there is a need for additional knowledge about the nature of the mind. However, this does not describe nor delineate methods that would make this possible.

Perspectives for the Natural Sciences

Currently in America, it is the remnants of our ancient minds that facilitate adaptational survival that threatens our very survival today. Our individual freedoms and rights allow dangerous afflicted people to go unnoticed and undetected until they commit acts of mayhem. Threats to life and property are just as great as those to the economy. It is important that science understand the mind.

It is my contention that the mind can only be understood by an integrated science. Modern science has made modest attempts to integrate its various disciplines. For instance, we have biochemistry, biological psychology, and math for the life sciences. However, these have not led to the answer. The question is, what is the best way to collectively use the sciences to study mind or the functioning of the entire nervous system?

Scientists consider physics to be what they term the fundamental science. Since all sciences come under the term natural, can the study of mind be successful without placing it within the confines of physics? Physics is the study of the physical world. Physics deals with such things as: motion, forces, energy, mass, and particles. Physicists, like other scientists, make observations, and their rules of observations should apply to all the natural sciences, and this is where the roadblock for understanding mind originated.

Physics as the fundamental science has its own long history of making changes to the rules and laws of observations from Claudius Ptolemy (second century A.D.) to Werner Heisenberg (1927). Because of and in spite of this continuous flux, progress has been the result. Strangely, it was Einstein's great achievement that separated physics from the rest of the natural sciences. Physicists needed a period of pause and separation from the other natural sciences. Bergmann (1992) made these statements:

> *More than in any other area of the natural sciences, any change in space and time concepts requires major adjustments in physics, because space and time are the scaffolding in terms of which physical processes must be described. Such adjustments are particularly called for in theories dealing with the fundamental aspects of particles and their interactions. The need for revision was not equally urgent in all areas of science. The exploration of the secrets of nature has always proceeded on many different levels of profundity and complexity.*
> *A living organism, for instance, is a system of enormous complexity, whose dynamics cannot be reduced to detailed description of all the interactions of its constituent atoms and molecules. No one would seriously suggest that biological research be differed until all the elementary processes which taken together, make up life are fully comprehended separately...*
> *Improved understanding at one level of complexity usually contributes to an understanding of adjacent levels but may be relatively unimportant at levels far removed.* (pp. 57-58)

Gamow (1988) a Russian physicist, while explaining the principle of time dilation, made these comments:

> *If a zero-time interval becomes larger than zero when viewed from a moving system, then a finite time difference between two events must increase when viewed from the same system. This is the famous time-dilation, or slowing of the clock (along with all other physical, chemical, and biological processes) when observed from a moving system.* (p.178)

This insight informs us that time dilation involves changes in physical, chemical, and biological processes. Isn't this what the brain and the nervous system do to create mind? Isn't our perceptual system that works with the brain a moving system? Gamow hinted that the time dilation effect might be proven to happen on a large scale.

So, can we bring physics to bear on the biological subject of mind or the workings of the total nervous system of man as physics exist in its present state of development? Do the other natural sciences have to wait on the science of physics to make new adjustments?

Chapter 2

MODELS OF MIND

Modern mind scientists come from a variety of disciplines. Sometimes researchers are confused as to who should do what research of the various aspects regarding mind research. They equate mind with the brain only and ascribe or confine all aspects of mental function to it. In addition, the mind is categorized as a complex system. This model of mind is a major problem, and its separation from the fundamental science of physics makes it impossible to define or to describe. Ironically, it was the US physicists of Einstein's era who led the life scientists astray. Einstein's new physics was so mind-boggling that neither he nor anyone else knew how to use it in all of its applications

This has resulted in mind scientists looking at brain structures and the cells they are made of for clues of mental functions. This disregards the fact that humans live in a physical world made up of physical forces such as gravity and atmospheric pressure, and that we have to respond to these stimuli on a continuous basis. So, I guess you could say that physicists threw away the baby with the bath water. Now you have mind scientists studying the brain for clues to the mind without connecting mind to the fundamental forces that we have to respond to. Goldman-Rakic (1993) demonstrates this tendency:

> *Until recently, the fundamental processes involved in such higher mental functions defied description in the mechanistic terms of science. Indeed, for the greater part of this century, neurobiologists often denied that such functions were accessible to scientific analysis or declared that they belonged strictly to the domain of psychology and philosophy. Within the past two decades, however, neuroscientists have made great advances in understanding the relation between cognitive processes and the anatomic organization of the brain. As a consequence, even global mental attributes such as thought and intentionality can now be meaningfully studied in the laboratory.*
>
> *The ultimate goal of that work is extraordinarily ambitious. Eventually researchers such as myself hope to be able to analyze higher mental functions in terms of the coordinated activation of neurons in various structures of the brain. It should also be possible to identify the cells that mediate the activity of those structures. Such research will help explain the origin of mind. It may also lead to more complete description of baffling mental disorders such as schizophrenia.* (p.67)

As you can see, this model is only concerned with the brain and what is going on inside the brain. The global aspects of mind are thought to be thought intentionally rather than the forces of physics. There is no consideration that mind might have an order that facilitates function. You could say that the absence of physics created a void leading to a lack of direction in mind research.

As I have read much psychology research that is pointing to the direction that further research should go in order to determine the origin and nature of mind, I am amazed that these hints have gone unnoticed by researchers. Current efforts seem determined to continue digging a deeper ditch rather than digging a new hole. One of the problems with these models is that they have limitations acting as blinders as you can only see or know what they allow.

Readjustments

Behavioral scientists have labeled every structure in the brain and determined their functions, and yet they can't understand how all this fit together in order to create mind. This alone it seems to suggest that maybe the brain by itself doesn't create mind. Maybe other structures and principles are operating in this process? In addition to physics, there may be evolutionary principles that are already known about man and at least apparent, if not completely understood, at present that nonetheless may give hints about the origin and nature of mind. Could it be that scientists have put the study of mind into a box that is too small and doesn't bring enough of existing knowledge to bear upon this subject?

The way we look at things can have severe consequences, early attempts to understand the Earth itself, and its place in the solar system, can be instructive. It was once believed that the Earth was flat, and this held serious consequences for those people in that part of the world as they were limited in travel, which negatively affected many aspects of everyday life. Everything changed when it was discovered that the Earth was actually round in shape. We must have accurate and complete fundamental knowledge of what makes up a system in its entirety before we can analyze it and use it. Once it was known that the Earth was round, it immediately removed fears of oceanic travel. Map making was enhanced ,and navigational knowledge and new instruments were invented as well as improvement in ship building and design. All this led to the discovery of new lands, people, cultures, food products, other goods, and trade.

Let us look further at science in particular physics, and let it be instructive in that physics has a long history of development and has had to make adjustments more than once. What I really want to emphasize is that in physics changes were and are being made, and that mind sciences are going to have to do the same! Let us start with Claudius Ptolemy in the second century A.D. His theory was that the Earth formed the center of the universe. He was wrong, and corrections had to be made.

With the passage of several hundred years and the critical observations of several men, Kepler, Brahe, and Copernicus, it was discovered that the sun rather than the Earth formed the center of planetary motion. Kepler's work paved the way for Newton to formulate the laws of planetary motion in 1609, and Newton's Principia was published in 1686, a period of 77 years. It was in 1905 that Einstein published his work on relativity. Werner Heisenberg presented the uncertainty principle in 1927. More than a millennium of work with great advances and with apparent contradictions.

Looking at the developmental process of physics as a model paradigm, the other sciences should consider that they might need to deliberately set out to see if they need to make adjustments, and there needs to be methods for doing such. Looking at this model, it is easy to see that the development of physics advanced in stages over a very long period of time. The goal now is to make adjustments quickly over a short period of time. In addition, there were circumstantial conditions at play, such as historical realities in each physicist's time, such as belief systems, social restraints, current level of development of physics, math and technology, and natural ability.

Neuroscientists such as Restak (1994) admits that neuroscience is about 60 years behind the Heisenberg principle of uncertainty when discussing the subjective nature of objective events (p.133). Here for the first time, we have a mind scientist admitting that the mind should be studied within the context of physics; however, he is also admitting that it is a long way off. Nonetheless, this is the start of an adjustment. Later on, I intend to demonstrate that we are going to have to go through Einstein's relativity first so that we can understand the origin of mind and its physical fundamentals. We cannot hope to understand mind while ignoring the fact man is a terrestrial creature and that his entire nervous system function as a unit in order to perceive events and to respond to them, and this is mind. Hawking (1996) stated, "the whole history of science has been the realization that events do not happen in a arbitrary manner, but that they reflect a certain order, which may or may not be divinely inspired" (p. 157). Man lives on planet Earth, which has a constant gravitational field and a fluctuating atmospheric pressure, and these fields interact with him constantly. With this being the case, why jump to structures and chemical processes that may be less important in the order that constitutes mind?

Complex Concepts and Modeling

As you may have noticed while reading, I have inserted bits of information gradually in order to facilitate a transition for the concept of mind. After looking at "the history of physics model," I hope that we can profit by condensing time so to speak and speed up our effort. We are going to move forward by proposing a specific order for using scientific concepts that already exist and expand the entire field of mind research to include all information that pertains to mind.

In order to create a bridge for this enhanced understanding, let us start with the fundamental or initial principles:

1 For studying mind, expand the concept of mind to include the entire nervous system.

2 Make models of as many concepts of psychology and neuroscience as necessary.

3 Make physics the fundamental science.

4 It is interesting to note that scientist have put together proper models for the study of any subject matter. However, researchers have strayed from them with less than desirable results. Myers and Jeeves (1987) presented the following concepts, the first their own, and the second, that of Polkinghore. The concepts of Myers and Jeeves represent levels of explanation. Each is incomplete but is a complement to each other, and all combined gives a more complete view of the subject. I consider this as a basic or fundamental model.

Figure 1. Integrative Explanation

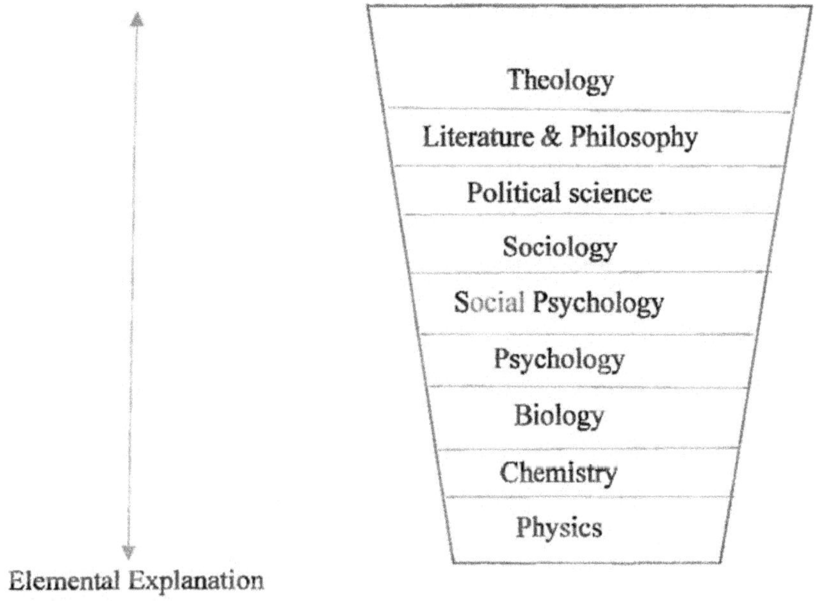

Looking at this model as a guide for research, you can see that physics is the fundamental science for explaining phenomenon (any observable fact or event that can be scientifically described). One can also see that mind scientists have primarily confined their efforts to chemistry and biology. In contrast, ancient science made use of physics to define mind, or "Nous" as they called it. According to James (1954), Anaxagoras, an ancient Geek philosopher, defined mind: "nous alone is self-moved, and is the cause of motion in everything in the universe and has supreme power over all things" (p.64). Mind was considered to be composed of fire atoms by these Greek philosophers, which I interpret as electricity. My point being that the ancients used physics to explain mind, and so will I. As this information has come to us, it may sound confusing; however, I can see that, if employed, a comprehensive approach with motion of man and his surroundings is the fundamental concept pertaining to mind. As I continue to present models, it will become apparent that I have rediscovered the same theory. I am sure that you have heard that there isn't anything new under the sun.

Next, I will use Polkinghore's concept and convert it to a basic mind model as Figure 2.

Figure 2. Ways of Perceiving Man

Ways of Perceiving Man
A species of homo-sapiens
A multi-layered unity
An open biochemical system in interaction with the environment
An aggregation of atoms

This model forces us, or should I say enables us, to look at mind in relation to its environment, to consider the unity of mind, and to compare the mind of man with other species from a more comprehensive perspective. Neuroscientists have made comparisons of the brains of other species and thus far haven't discovered any reason why man alone has what is considered mind. However, they are only comparing one nervous system to another and not looking at all the other differences that separate us. This model also allows us to deal with issues that we have the capacity to understand with our current knowledge base and level of technology and allows us to fully develop the mind concept in totality, which keeps us from going astray as well.

As we continue to use more and more advanced models in our search to understand mind, we must discuss the contribution of John Hughlings Jackson. Jackson definitely set a foundation for understanding mind in more than one way. First, he gave an accurate description of the mind as he went. Second, he described what I believe is the major component of mind and put me on the right path in my search of mind. If not for Jackson and his explanation of his observations, I would not have been able to make mine. Jackson was said to have had a great interest in epilepsy, which he thought was the key to the functioning of the brain. Jackson noticed a connection between mind and motion, but he didn't go any further than this basic idea. Even though his philosophy managed to steer mind research in the direction of brain as mind with the neglecting of the rest of the nervous system and physics, his piece of the puzzle was huge. Mind scientist had great interest in Jackson's theory that mind was the result of motor responses to sensory stimuli, but they didn't make connections to the science of motion.

I am going to make a model diagram of Jackson's concept of mind so that I can use it as a bridge to a greater understanding and as a visual tool for comparison.

Figure 3. Jackson's Model

Brain is mind-a pyramid of increasingly complex functions

Cerebral hemisphere prefrontal lobe in control under ordinary conditions

a Rational thought

b Conscious intentions

c Hold in check the more instinctual and automatic responses of the lower levels

Lower levels that well upon injury to higher centers

d Located beneath the hemispheres

e Brain stem

f Spinal cord

Sensory-motor units responsible for all mental functions

Modeling the Entire Nervous System

Scientists study physical fossils to help us understand the past and gain knowledge in order to help us understand the present. I am arranging these models to do the same. Hopefully, this model approach will have multiple benefits: lead us in the proper direction; help us formulate the proper concepts; and put everything in its proper order. Eventually, I hope to demonstrate the evolutionary nature of mind that the mind also has fossils and that they are still a part of us. In fact, we are walking around with them; we are born with them. They are components of our modern minds as parts of mental constructs or behavioral sets that were once adaptive, but can be rearranged within the total mind construct and become maladaptive. The Bible in the book of Revelation Chapter 4, verse 8 mentions living creatures that were full of eyes around and within.

As scientists today, we have more eyes as a result of accumulated knowledge. The flip side is that we do have more to look at and that is the negative aspect of scientific research as sometimes we are blinded by so much information or fail to put all information in the right context. This is why I am making use of models so that our search will have eyes and that we can have the necessary guiding principles to keep on a track that will include all of the proper and necessary information. This allows us to have a psychology for research that expands the rules, methods, and techniques, thus bridging access to all available knowledge and allowing us the vision to put the various concepts in their proper place. Unfortunately, there are no guarantees inherit within the arena of scientific research that science left unguided will always come to the truth. So, it is within this frame work that we will proceed model upon model as a type of synthetic research mind.

I am hopeful that this readjustment in perspective and process will enhance understanding of mind and create a means of progress. In this regards, I am proposing my initial mind model (Figure 4.) The elemental stage of this model includes the brain and the nervous system, including the vestibular apparatus. This is a major departure from all of the concepts of science on the subject of mind. Current literature doesn't talk about the vestibular system in connections with mind theories. This is the bridge that allows us to connect mind with the physical forces of nature. The vestibular apparatus is connected to the brain and the rest of the nervous system as well, as to the fundamental forces of nature, and without it, we are not vertical creatures.

Figure 4. Leonard Wilson's Elemental Mind Model Mind

The vestibular system – major organ facilitating mind
The brain – dependent upon input stimulation from vestibular in order to manifest mind and maintenance of its' structures
The entire nervous system – necessary for the total functioning of vestibular
The environment – fundamental forces necessary for vertical orientation that provides structure for mind from activated, functional vestibular

Figure 5. Leonard Wilson's Intermediate Model

Mind Vestibular System Operating Principals Relativity   Basic Math – Advanced Math   Uncertainty Principle
Weighted in regards to health and size of vestibular apparatus and environmental conditions

Figure 6. Leonard Wilson's Complete Model

Mind
Vestibular system/Entire nervous system
Jacksonian March
Heart rate
Autistic spectrum
Rem\Awake cycle
Behavioral: daily/seasonal/biological relativity (asymmetry)
Psychological Safety/Stress
Uncertainty principle (perceptual properties)
Relativity – horizontal vs. vertical orientation
Electrical storms/chemistry in motion
Vestibular-brain-neuron structures-mass
ENVIRONMENT – 1G – gravity, sunlight variation, atmospheric pressure, and manmade artificial forces.

So, now we have transitioned from the simple concept of mind as brain or vice versa and back to the very ancient idea that mind consists of the entire solar system, with its motion, the Earth with its environment, and a human body that connects to the entire system via the vestibular system. The vestibular is a biological organ that functions by using physics and mathematics. This will be discussed in greater detail later. When we include the vestibular, brain, and the entire nervous system, a whole new picture begins to emerge, and we encounter a totally different understanding of mind. I would like to emphasize that mind itself is a very large operation, functioning using physics and math at different levels in order to perform different behaviors.

Mind requires a multilevel of sequence to be functional for a mind to function. Yes, the brain is a necessary component of mind for consciousness to exist. For the brain to perform its functions all of its necessary structures must have the proper components, size, shape and neurochemistry and electrical activity. This can't happen without proper stimulation from the vestibular and proper nutrition connected to proper growth cycles and movement science. Hopefully, the ground work is being set so that observations made and reported by great scientists such as Hawkins and Mlodinow (2010) can be seen or placed in the proper context, such as the following:

> *Though we feel that we can choose what we do, our understanding of the molecular basis of biology shows that biological processes are governed by the laws of physics and chemistry and therefore are as determined as the orbits of the planets. Recent experiments in neuroscience support the view that it is our physical brain, following the known laws of science, that determines our actions, and not some agency that exists outside those laws. For example, a study of patients undergoing awake brain surgery, found that by electrically stimulating the appropriate regions of the brain, one could create in the patient the desire to move the hand, arm, or foot, or to move the lips and talk. It is hard to imagine how free will can operate if our behavior is determined by physical law, so it seems that we are no more than biological machines and that free will is just an illusion.* (pp. 31-31)

This is a very important point as well as a point of departure from my perspective. We are biological machines; however, because of our vestibular system, we have the capacity of will. Man, in his early stage, was without a will and autistic. As man became more upright over time with an improvement in movement, the vestibular had enough computing power for subjective thought and a will. More on this later. It is the vestibular that ultimately controls motion, not the brain; and yes, the vestibular provides the electrical stimulation for the brain to do what it does. The vestibular with its tremendous computational power and its control of the brain and the rest of the nervous system is mind. The vestibular stimulates brain structures electrically, and this is why even external currents can also stimulate brain structures as this is how the vestibular stimulates the brain. We will continue to develop this point. However, humans are subject to various stimuli that can cause the mind to malfunction. Space biologists noted that astronauts experience mental problems when they are subjected to micro-gravity, and they called this stress "altered sensory stimulus states." I have included a list:

- Micro gravity
- Non uniform motion or acceleration
- Physically threatening stimuli
- Motion across time zones
- Rapid change in blue light stimulus
- Exposure to two accelerations
- Change in atmospheric pressure

- Change in atmospheric pressure from the motion of weather systems
- Natural changes in blue light due to movement of the earth and sun and the earth's atmosphere

Supersensitive vestibular responds by altering physical and mental states.

Also, since our mind models have led us to the conclusion that the solar system and its motion is the foundation of what we call mind, taken to the next level, our entire solar system is a machine designed to operates so that a biological mind can and does exist. In other words, our solar system is no accident, and neither is life. However, from this perspective, mind is limited. Restak (1994) proposed that the brain is modular, and this facilitates mind and to me this also suggest limitations as only so many modules exists and I call them mind templates. In addition, I would like to note that humans have only so many thoughts within a certain number of domains and only for so many of hours of the day, and this 24-hour cycle is a result of our solar system. We have only so many experiences in a day, season, or in a year, and these repeat over what we consider short life spans. So, our solar system and environment together provide a potential for a biological mind; they also limit that mind.

Chapter 3

THE MIND: A COMPLEX SYSTEM

Up to this point we have presented models that would allow us to generate more information about mind. As a result, it became apparent that mind is much more that a human brain. Mind requires much more that a brain, even though, there would be no mind without it. Mind requires an entire nervous system, as discussed by Jackson (1993) as he quoted professor J. P. Guildford:

> *"The mind," says Professor J. P. Guilford, "if it has any scientific meaning at all, is the functioning of the brain. But since the brain cannot function alone, but only in connection with nerves. receptors and effectors, mind is better described as the functioning of the entire nervous system."* (p.48)

It also requires a vestibular, 1-G gravity, and a 24-hour night/day cycle with variation in light intensities, atmospheric pressure, upright posture with a bipedal gait, and REM sleep. Everyone knows that humans don't fare well in micro gravity. There is abundant space biology research that attests to the negative effects both physical and mental acquired by astronauts who spend time in microgravity. The vestibular system malfunctions under such conditions, proving that our solar system, including the planet Earth, are perfectly designed to our bodies and thus also mind. Mind ultimately is a terrestrial function, and it is in the form of a human body connected to the solar system, Earth, and its environment. Stated in another way mind is a multilayered unity or harmony, and exists only because of and within the boundaries of our solar system. How can we understand mind and its various behavioral manifestations, and cure and prevent mental illness, if we only look at the brain with its structures, chemicals, and electrical properties?

Would it not be beneficial to consider the mechanics of the mind as it relates to the environment and the role of other organs and functions of the body in regards to their contribution to mind?

Mind and the Vestibular

Mind scientists and others who study physiology and compare species have long considered what difference in brain structures of the various species could account for the human mind. This effort has resulted in much speculation, but no certain answer. Once again, even though there is a tremendous body of knowledge concerning the human body and the human condition, without modeling, scientists don't know that they need to consider additional information, much less what that information should be. First of all, when comparing species for a clue for understanding mind, the thing to consider is motion. Remember the Jackson Model: it was the sensory motor units that were responsible for all thoughts. Human locomotion scientists had this to say:

> *The fact that we humans are bipeds and locomote over the ground with one foot in contact (walking) no feet in contact (running), or both feet in contact (standing) creates a major challenge to our balance control system. Because two-thirds of our body mass is located two-thirds of body height above the ground we are an inherently unstable system unless a control system is continuously acting.* (Abstract)

This comparative difference of the extreme demand on the human vestibular system for balance and motion facilitates thoughts or mind. This same level of sensory motor control by the vestibular allows for human speech. We can rightfully state that mind is a multilevel, multicomponent, complex harmony. Even though the vestibular system in lower animals possess extreme computational abilities, the system capacity in man is many orders greater. Man, with his body design, vertical orientation, and bipedal locomotion separates him from the rest of the animal kingdom and fits a vestibular controlled brain and nervous system capable of mind.

With this information, we could further conclude that our solar system is man, or mind. This sounds like a modern version of an ancient theory. In addition, this information agrees with the idea that man is on Earth for a purpose, and currently, he is at the last level or phase of a gigantic machine that we call mind. Lets' take a closer look at the unmoved mover or what we call the vestibular system.

The Vestibular System

I am going to give a general overview of this system, enough information to make my point. I would advise those readers who would like more information to look up this system in physiology texts and the internet. According to Moyer (2014), this sense is one of the first to develop in a growing fetus and is well developed by five months. According to Moyer:

> *This system is very important to a child's early development. Its role is to relay information to the brain as to where a person is in space, as related to gravity, whether they are moving or still, if they are moving how quickly, and in what direction. The vestibular system gathers that information from a set of fluid filled canals and a sac-like structure in the inner ear. These structures respond to movement, change in direction, change of head position, and gravitational pull.*

The vestibular apparatus is a tiny organ located within the inner ear. Without this system, we can by no means function or live on planet Earth. Mind is a machine consisting of redundant multi-level movement. The planets moving, neurotransmitters moving, electrical current moving, and the body moving in space.

Due to its importance to mind, I am going to quote some additional information. Hopefully in so doing I will be setting the ground work for a clearer understanding of mental illness later on. I would like to begin by quoting an internet article, "Vestibular Dysfunction" (2014):

> *It's amazing that something so small can detect even the slightest bit of movement we make. In fact, all sensations are interpreted with reference to the vestibular system. The nervous system depends on this working properly in order to process sensory input from all of our senses. If there is vestibular dysfunction, then all other aspects of the nervous system will fail to function accurately as well. (p.1) I would like to list three very important statements made by Parker (2012) on the vestibular:*
>
> - *First sensory system that will completely develop by six months after conception*
> - *Sensory system that is thought to have the most important influence on the other sensory systems and on the ability to function in everyday life*
> - *Directly or indirectly this system influences almost everything we do. It is the unifying system in our brain that modifies and coordinates information that is received from other systems. The vestibular system tells each sensation where and when it should go or stop.* (p.1)

Looking at it this way, it seems that the brain and the rest of the nervous system developed as instruments or vehicles to facilitate the capacity inherit within the vestibular system, and as such, they are extensions of the vestibular system. In psychology and physiology textbooks, the vestibular system is not given much attention nor importance. In neuroscience books devoted to mind, it isn't mentioned even when functions that its controls are under consideration. For instance, scientists have discovered that a person's personality is preformed before birth as a property of heart rate, and it is well known that mental illnesses are associated with a smaller than normal hippocampus. Both of these are under the control of the vestibular. But mind scientists stop short of considering the vestibular; in my opinion, a serious hinderance. It appears that a complete reevaluation is in order in regard to the nature and function of the human nervous system and a revision of our psychology and physiology texts.

To further illustrate this point, I would like to cite an article from Wikipedia encyclopedia, "Vestibular System" (2014), in which it said this about vestibular processing:

Signals from the vestibular system also project to the cerebellum (where they are used to keep VOR effective, a task usually referred to as learning or adaption) and to different areas in the cortex, the projections to the cortex are spread out over different areas, and their implications are currently not clearly understood. (p.4)

Hopefully, using this unique approach to mind, I am making progressively clearer statements about the subject. So far, we have shown that mind isn't the product of a local system, such as the human brain alone, but it is a global entity as the result of the entire solar system in motion, a gigantic machine with a vestibular system interfacing with the external and internal environments of a human body resulting in what we call mind.

Let us take an even closer look at the vestibular system by looking at another article from the internet called "Brain Breakthrough" (2014) as it connects the vestibular to mind:

The Vestibular System and the Brain

The vestibular system in the brain allows us to stand upright, maintain balance, move through space, and process visual information about the environment while we are in motion. The vestibular system is a surprisingly complex sensory system, integrating information from:

- *The vestibular organs in the inner ear*
- *The eyes*
- *The neck and shoulder muscles*
- *The finger tips and palms of the hand*
- *The pressor on the sole of the feet*
- *Hip and leg joints*
- *Jaw muscles and facial muscles*

The vestibular system is perhaps the most basic of all the sensory systems. The system develops early in utero and through the many connections with the rest of the brain, it is believed to form the foundation of many other functions. The vestibular system coordinates body movements, maintains balance and equilibrium, and helps children develop normal muscle tone. It plays a significant role in the development of language, such that children with vestibular dysfunction may also have auditory-language processing problems.

The vestibular system influences motor control and motor planning that are necessary to use the fine muscles in the throat, tongue, lips, and jaw to produce intelligible speech. Because the vestibular system is crucial for effective auditor processing, the child with vestibular dysfunction frequently develops problems with language. In addition, other input mechanisms that affect and are affected by the vestibular functioning are the hip and leg joints impacting large motor coordination, hands and fingers affecting fine motor skills, and the pressor on the soles of the feet giving a sense of stability and balance in space.

It's important to realize that the vestibular, auditory and visual systems work together as they process sensations of movement and sound and light. These sensations are closely entwined, as they all begin to be processed and are influenced by receptors of the ear. Vision is an important component of the vestibular system. About twenty percent of visual neurons respond to vestibular stimulation (eq. When spinning, head shaking or rocking). The auditory system is also highly involved in vestibular functions. The vestibular and auditory nerves join in the auditory canal and become the eight cranial nerve of the brain. Anything that disrupts auditory information can also affect vestibular function.

One system feeds the others information that allows us to think, move, comprehend and interpret the world around us. We call this process of balancing and improving the communication between input systems, sensory integration. (p.1)

The human vestibular system, which is more powerful and sensitive than any other species, makes it possible for enough sensory integration that thought is the final product. Don't forget that it is the numerous stimulations from the solar system, the Earth's 1-G gravitational field, sunlight, motion and our atmosphere, and our own motion in space that causes this system to work. Next, let us look at mind at the individual level. What is the genesis of mind in each person?

Heartrate and Mind

Every living human being has a mind of their own. Even though, there are personalities that people are grouped in, no one thinks exactly the same, and sometimes people are diagnosed with abnormal minds and personalities.

I am going to define personality as our basic or fundamental mental frame of reference for interacting with our environment. Campbell (1940) defined personality as follows:

The characteristic, and to some extent predictable, behavior-response patterns that each person evolves both consciously and unconsciously, as his style of life ... The personality functions to maintain a stable, reciprocal relationship between the person and his environment: it is thus a composite of the ego defenses, the autoplastic and the alloplastic maneuvers, that are automatically and customarily employed to maintain intrapsychic stability. (p.455)

We are now going to see why the vestibular system is the first sensory system to develop in the fetus.

Yes, the vestibular system is at work creating our foundational personalities (minds) by effecting heart rate differences in the unborn. Kagan (1999) while doing research to determine the biological predictors of temperament found the following:

> *When the mothers of the infants were pregnant, they came to our laboratory three weeks before their expected date of delivery. Each woman lay on a recliner as we measured her heart rate. A computer program separated the mother's heart rate, which is low, from the higher heart rate of the fetus. What we found was that the fetus whose heart rate was 140 beats per minutes or higher had a high probability of testing later as a high-reactive four-month-old. (Some who had heart rates of less than 140 beats per minutes did become high-reactive, but the probability of that occurring was lower.) We also measured the infants heart rates when they were two weeks old. We visited the home, waited for the infant to fall asleep, and then measured heart rate at two different times-once when the sleeping infant was supine and again when we asked the mother to pick her sleeping baby up and hold the baby over her shoulder in an erect posture. The erect posture provokes activity in the sympathetic nervous system, raising heart rate and blood pressure, which is why we can get out of bed in the morning without passing out. We found no difference in supine heart rate between the infants who would later become low reactive. However, the two-week-old infants who tested as high reactive three and a half months later had significantly higher heart rates while sleeping in an erect posture than did those who later tested as low-reactive.* (p.42)

The high reactive infants were fearful and shy and inhibited. The low reactive infants were uninhibited. It is important to note that mind states are preformed before birth, and the organ that controls balance and heart rate is the major player. Considering this, it appears that mind is the purpose of a human existence. Purpose and a design for the manifestation of mind. As soon as a human starts to form in the womb, it develops the system that lays the ground work for the continuation of what will become mind. All of the elements working in repetitive sequence produce mind. If for any reason just one of the elements that work together to create this giant machine that we call mind is altered then so is the mind. I can't help but wonder, what are the probabilities of these sequences happening by mere chance? I haven't covered all of the sequences yet. I would also like to state that my current understanding only allows me to start with the solar system at the moment. Nonetheless, contemplating that we have this giant solar system in motion interacting with planet Earth and human beings in such a precise way, the end result is mind is mind-boggling!

The Jacksonian March

In the last section, we talked about mind as it begins to develop at the human level after conception and just before and immediately after birth as the result of a biological system developing first known as the vestibular system. Matter comes together to exist in this part of our universe known to us as the solar system and then forms into planets with necessary shapes and masses, creating the necessary orbits and the precise positions in regard to each other. Biological life existing on Earth, which has the proper elements in the proper amounts at the proper time or sequence and the human body, is in sequence with all of this with the result being biological life and what we call mind. Too many sequences to be accidental!

Yes, you guessed it, there are still sequences to cover in regards to mind. Also, all biological creatures are programmed with a clock known as the circadian cycle in which body functions are coordinated with variations in sunlight intensity over a 24-hour cycle in a sequential manner. But how do we get from Kagan's basic personality to mind and will?

We discussed earlier about the allometric design of the human body and the tremendous demands put on the human vestibular system to maintain balance in a vertical position with a unique mode of locomotion. What is the sequence that leads from a basic personality to mind and will, and how is it related to a vertical frame of reference?

As humans, we are designed to observe events in the physical realm. Yet it is surprising how little attention we pay to everyday events, even though, we are already equipped to do so. Strangely, as an outstanding observer of life events, I didn't pay attention to one event that was presented to me over and over again.

I was 14 years old the first time I encountered this particular event. My history teacher, who had been my teacher in seventh grade, PE teacher, football and track coach had a dramatic seizure while teaching world history. Later, with a long history in the criminal justice system, I would be on the scene numerous times as inmates had seizures. I never became interested in this subject until I decided to do an advance study of mind.

It is my belief based on my life experience that we can find solutions to some of life's most complex problems by making close, comprehensive observations of natural events in our natural environment. It was while I was reading literature on seizures, a sudden light came on.

John Jackson was interested in seizures and one type in particular that is named after him, the "Jacksonian March." Restak (1994) describes this seizure in detail:

> *In focal epilepsy, the seizure will progress in a definite order or – "Jacksonian March" involving the face first, then the hand and arm, then the leg on the same side and finally a generalized seizure that catapults the patient to the ground. During one of these seizures the patient often experiences a disturbing and inexplicable temporary a loss of will to move. Even more frightening and disturbing are experiences of an un explainable fear and dread. At such times the patient feels distant from himself, strangely altered and beyond conscious control. With the end of the seizure, the patient returns to his or her normal state.*

What do we learn from this natural phenomenon? Jackson concluded after careful observation that mind was the result of motor-sensory units, and that motion and mind were the result of the same processes. This was in the 1870s, and Jackson was concerned with the brain. From my perspective, this is a large revelation in plain sight.

First, the final result of our vestibular system integrated with the entire nervous system with its enhanced computational abilities, which are required to maintain vertical balance with unique physical dimensions of a human body and the stress imposed from bipedal locomotion endows this system with a unique property that we call mind.

Just as Jackson made observations, so will I. It appears that to change types of consciousness or mind states that go from a lateral or supine position to a vertical one is necessary first step. When the vestibular is occupied with vertical muscle tone and motion, mind is also present. However, to make a change in mind states the vestibular has to go through a sequence of letting go of muscle tone for vertical orientation. During our 24-hour circadian cycle, we humans switch from a vertical to supine position daily, and we cycle through mind states. The fact that a person loses both will and consciousness by falling during the Jacksonian March seems to be a portrait of a body going through a reverse process and that the standing up vertical is the process that turns on mind.

Even though I don't mean to offend anyone, this sequence in loss of muscle tone may mean that it took a long time for mind to develop; it may not have an instantaneous process. Man may not have always been a vertical creature, and he may have improved posture over time, just as indicated by the models of man in prehistory by evolutionary theorist in which man is presented as gradually standing in a vertical posture in incremental advances.

This idea has been suggested by men of science such as Halstead (1989), in which he states:

> *With the gradual improvement in gate and posture of land vertebrates. there was a marked improvement in the sense of balance and muscular co-ordi-*

> *nation, particularly among members of the group that took to the air. This is reflected in the enormous increase in the cerebellum, which in primitive vertebrates was only a bundle of nerve fibers at the front margin of the hind-brain.* (p.23)

I would also like to note that in REM behavior disorder that a seizure can cause a person to become vertical and respond to a dream state. I mention this only to say that mind states could be attached to incremental development along with posture. It is currently believed by some mind scientists that all of the sleep states represent a continuum of mind.

So, in the final stage, mind is a 24-hour entity that moves along a continuum of awareness states and then reverses itself over and over, and this cycle is under the control of our environment, specifically variation in sunlight intensity. Mind in its manifestation is limited in more than one way. There are only so many sets of mind modules that contain so many thoughts, and you can only possess your maximum mind for a few hours of the day. We have all these systems working together forming a multicomponent device, all part of a gigantic machine. The universe that creates matter that forms the elements became the planets of our solar system, which includes planet Earth, where everything is in place so that mind could manifest. Who designed this machine for the purpose of mind?

Another Continuum: The Autistic Spectrum

We must mention the fact that some mind scientists believe that all the stages of sleep represent mind at various levels, a continuum of mind states. Evolutionary scientist Jones (1994) stated That, "Everyone is a living fossil carry within themselves a record which goes back to the beginning of humanity and far beyond" (p.15). An earlier scientist named Haeckel proposed the Biogenetic Law, which states: "Ontogeny is a brief and condensed recapitulation of phylogeny." This law means that the individual organism in its embryological development passes through stages in which it duplicates in an abbreviated form, the history of the species. The human embryo is said to go through the form of early life forms in succession before it takes on human form. If we are caring around in our genes the history of our entire past, then all our mind states are also within us and can be operative under the right circumstances.

After doing a comprehensive study of the vestibular system, I believe that early man was much smaller than modern man. In fact, I believe that early man is still with us today in the people that we call pygmy. King (1994) declares, "the ancient Egyptian left an indelible proof in their mythology of their decent from the first human, which was the pygmy" (p.18). These people are reported to be hunter gathers, and as such, they remain in their original mode of life. With this mode of industry, it would have been advantageous to have mind states such as map calculating, calendar calculating, and math calculating for basic survival. If the Earth's climate and atmosphere were different in the early stage of man, and some believe it was, then it would have been crucial to know when and where food was found before without any external and environmental clues. People who have these calculating abilities today are known as autistic servants. In addition, I can't imagine that there was much time for socializing behavior, as survival would require great effort.

I also believe that as we look closer at the seizure called Jacksonian March and note what was said about the mental aspects of this phenomenon, it gives additional clues about man's original mind states. Lets' take a step backward and look at just the mind state of the Jacksonian March:

> *During one of these seizures the patient often experiences a disturbing and inexplicable temporary loss of the will to move. Even more frightening and disturbing are experiences of an unexplainable fear and dread. At such times the patient feels distant from himself, strangely altered, the world and things around him threatening and ominous. All actions are temporarily impossible, and the "higher" mental functions of reason, memory, and emotions are altered and beyond conscious control. With the end of the seizure, the patient returns to his or her normal state.*

It seems that this phenomenon describes not only the sequence that jump starts mind, but also the historical sequence of mind states. Compare what Grandin (1995) say about the emotional state of some autistics, "some autistics have a greater state of fear and panic a state of hyper arousal as if being pursued by a dangerous attacker in a world of total chaos" (p59). Our past, it appears, is still with us both internally and externally; one needs only to look closely, carefully, and correctly.

Early man was at the autistic end of the spectrum. Man, in the beginning was, small and very different mentally than he is now, even though larger numbers of people are being born at the autistic end of the spectrum currently, and there are reasons for this to be discussed later.

As man began to grow and his movement improved, additional, more advanced mind states developed. As time passed, more and more people moved towards the higher functioning end of the spectrum. Our thought process is related to our behavior pattern, and in fact, they are limited by what we do for survival. By this standard, all living creatures on this planet can be considered as having a mind or as being a mind. With this being the case, then every behavior we have displayed throughout our entire evolutionary history is stored in our nervous system and under certain conditions can be retrieved and displayed. After years of evolution, we now have what is known as the autism spectrum as a foundational part of the human condition. Since autism exists with elements of depression, anxiety, and schizophrenia, which may have given primitive man some advantage survival traits that are very maladaptive in our modern society, it can surface as mental disabilities today as our vestibular system may try to use them as solutions to "altered sensory stimulus states."

According to the Center for Disease Control there are three different types of ASD:

1. Autism Spectrum Disorder (also called "classic" autism). This is what most people think of when hearing the word "autism." People with autism usually have significant language delays, social and communications challenges, and behaviors and interests different from those without autism. Many people with autism also have intellectual disabilities.
2. Asperger Syndrome. People with Asperger Syndrome usually function at a higher level of the autism spectrum. They might have social challenges and different behaviors and interests. However, they typically do not have problems with language or have intellectual disabilities.
3. Pervasive Developmental Disorders. Not otherwise specified, also called "atypical autism." People who meet some of the criteria for autism or Asperger Syndrome, but not all may be diagnosed with PDD. People with PDD usually have fewer behaviors than those with autism. The behaviors might cause only social and communication challenges. (N.P.)

This information has led me to propose a more concrete model of mind in which mind becomes an extension of autism: Human consciousness in its totality is an extended version of autism. Even what we call normal is nothing but autism with what we understand today as a brain with the appropriate volume and functional vestibular system. Mind is an autistic system that has evolved. It is a hierarchical and redundant system derived from a hierarchical and redundant nervous system.

1. Autism Spectrum
   Autism Spectrum Disorders
   - Classic Autism
   - Asperger Syndrome
   - Pervasive Developmental Disorders
2. Vestibular Complex Neurological Syndromes
   - Personality Disorders
   - Mental Illness
3. Normal Mind Templates
   Our evolved vestibular system working with full volume nervous system. Autistic elements compiled and arranged together in ways facilitating life.

4. Gifted
    - Athletically
    - Academically
5. Super Human
    - Gifted, both physical and mentally with potential for greater lifespans

COMPLEX NEUROLOGICAL SYNDROME-MIND TEMPLATES

Example:

Vestibular dysfunction – complex neurological syndrome – mind template
Reflex deficits
Spatial memory deficits
Autonomic disorders
Anxiety disorders
Depression

In closing out this section on autism, I would like to say that mind states (templates) are like viruses, and they can be changed (manipulated) or evolve to a certain degree all under the control of the vestibular system to meet environmental adaptive demands. However, stress in any form, psychological or environmental, can cause the vestibular system to malfunction, and this includes a damaged vestibular. Our society is producing more and more humans who function at the autistic end of the spectrum. Since the true nature of mind wasn't understood until now, no measures were taken to prevent this occurrence and as a result many autistic and savants have existed undetected.

Even a healthy vestibular can only function with information that exists in the system, and when faced with difficulty, it responds the same way with templates that may not be appropriate leading to disorders. As society becomes more sophisticated, more and more people are in reverse evolution. The above mind model indicates that mind in a human form is very plastic, and since the mind is the result of evolution, it can change forms quickly but always within a restricted range of autism, as no matter how far we are along the normal range, elements of autism are present in some form.

In this regard, I have encountered numerous people who were undiagnosed savants and were autistic.

The numbers are much higher than publicized. Looking at a psychiatric definition of autism, you can see how many autistics can avoid diagnosis simply because no one get close enough to know them. Pay close attention to this definition by Campbell (1981) as follows:

A form of thinking, more or less genuinely of a subjective character: if objective material enters, it is given subjective meaning and emphasis. Autism generally implies that the material is derived from the subject himself, appearing in the nature of daydreams, phantasies, delusions, hallucinations, etc. The content of thought, in other words, is largely endogenous. In classical instances of autistic thinking, such as occurs in schizophrenia, the unconscious sphere makes the largest contribution to autism.

Autism, dereism and introversion are closely allied to one another. Some authorities speak of the autistic temperament, meaning by the introverted, retiring type that shrinks from all contact with life.

Autism, when it is a pervasive and generalized dereistic life approach, is one of the fundamental symptoms of the schizophrenias (Bleuler).

In the autistic life approach, the -Me" predominates, often to the exclusion of the "Not-Me", external reality gradually loses more and more of its significance, and the patient is attuned to and guided only by the inner workings of his being. He is totally selfish in the most literal (and yet non-pejorative) sense of the terms, for he is unable to turn his energies onto objects outside himself. He exaggerates the importance of inner physical sensations as well his own ideas and emotions. He becomes involved in pseudo-philosophical speculations and has no time for mixing with his peers. He feels different from others complains that he has never realized his potentialities, and is concerned with establishing identity for himself (p.66).

I have lived with several people who fit this description exactly, except they had additional symptoms of extreme irritability and were very abusive to others. I could only live with them for several months, and their own family avoided them. I've worked in institutions where a person could lose their life if someone thought they had a larger piece of cornbread or chicken on the other persons plate compared to theirs.

Now we have come to a point where we can see that man or mind (and we could even call our entire solar system man or mind) is part of a giant machine where motion and force are key components connected to a biological component via the vestibular system that is the most powerful vestibular system in the animal kingdom due to the fact that it has to keep a vertical position even during bipedal motion. With the body of man having proportionate dimensions, that makes it extremely difficult to do so, and this extra computing power equips man with the final properties of the entire nervous system, which is mind. Nonetheless, this system as it exists can only be described as limited. It can only think so many thoughts and for so many hours during a day, and if anything changes at any sequence point in the system, it breaks down. When this system is confronted with what space biologists call "altered sensory stimulus state," which I expanded upon further understanding to that which is physically or psychologically threatening, or is confronted with nonuniform motion or acceleration, it responds habitually by going back to its original mind state or some distorted version of autism, depression, anxiety, and schizophrenia, etc.

Additional proof that the final state of the Jacksonian March is associated with autism and that this is all under the control of the vestibular is that these are the same depersonalization and derealization systems discussed by Smith (et al., 2013) in their work on vestibular impaired causes of personality changes:

> *Unlike the spatial memory deficits that have been documented in humans with bilateral vestibular loss (e.g. Brandtel., 2005), it is difficult if not impossible to disassociate the depersonalization and derealization symptoms such as feeling "spaced out", "body feels strange" and not "being in control of self', from symptoms that are a direct consequence of the impairment of the VORs and VSRs, such as oscillopsia and ataxia. (NP)*
>
> *In order to continue our understanding on this point lets continue quoting Smith (2013):*
>
> *Although the immediate and most obvious effects of poor vestibular function are oscillopsia and ataxia (see Curthoys and Halmagy; 1995 for a review), vestibular dysfunction results in a complex neurological syndrome characterized not only by reflex deficits, but also by spatial memory deficits, autonomic and anxiety disorders (see Smith et al., 2010 and Gurvich et al., 2013, in press, for reviews). In one of the first clinical studies, Grimm et al). (1989) described patients with a perilymph fistula syndrome (a rupture in the labyrinth, resulting in leakage of perilymphatic fluid) who experienced not only vestibular symptoms (e.g., positional vertigo) but also a variety of cognitive and emotional symptoms, including memory and attention deficits, anxiety and depression. (N.P.)*

Is there any room for doubt? We humans are in a state of confinement. We are all subjected to the same external stimuli, and everyone's nervous system has the same program. The program can only respond the same way, and it does. However, the system does seem to suggest clues to a solution. Maybe, we are not doomed!

Chapter 4

EVOLUTION, THEOLOGY, AND MIND

Evolution is a very sensitive subject and stirs the emotions of those on both side of the issue. The major arguments against this theory and the major problem proving it is that, as it is presented, no one can actually observe it happening in humans. So, I intend to get around these difficulties by sticking to my plan for this book by observing facts or processes that can be seen in their entirety or that are already known, even though they may be being put together for the first time.

In the worlds of weightlifting and body building, we see dramatic changes in both form and function of a human body after several years of consistent and dedicated training. I went two steps further in my work and proved that both size, foot speed, and athletic performance could be drastically changed in a prepubescent child and without the aid of drugs. In order to develop foot speed in a person of any age, I developed a system that I call the asymmetric grid. In this system, speed increases in small increments for a period of time and then suddenly a large jump in performance at once. Of course, the two giants of the theory of evolution proposed contrasting views as to the mechanics of the process. Charles Darwin was the proponent of a long, gradual process with many small increments of change, and D'Arcy Wentworth Thompson proposed sudden wholesale changes in form and function of a species. My grid program shows that a system involved in making changes in a biological system can involve both forms of change.

The system only takes a few months or years to work and observe. My reason for this short discussion on evolution is to discuss possible solutions to the problem presented due to the nature of mind and the crises facing the United States as a result of it. I believe that our current science and our Bible offer solutions. Unlike most, I view what is called ancient Greek philosophy and the Bible as serious advanced works of science. As can be seen, my exposition of mind exactly corresponds with the ancient Greek definition with all the blanks filled in. Later, you will see that my motion science is the same ancient science in the Bible.

The Bible and Science of Motion

We have seen that the ancient Greek philosophers included motion in their definition of mind. I have shown that mind is a direct product of the unique motion of man different from all other species. Our vestibular system is concerned with motion, and as such, it is an observation system of motion. Saxon (1993) stated, "the coordinate system an observer uses to measures positions, velocities, and accelerations is called the observes reference frame. Every observer has a reference frame attached to his head" (p.18). Saxon's book was a book on the subject of physics and as such, didn't make a reference to the fact that in our heads is the vestibular system, and it is concerned with motion.

Let us look at the vestibular from the perspective of motion, according to Carlson (1988) description:

> *The vestibular system has two components: The vestibular sacs and the semicircular canals. They represent the second and third components of the bony labyrinths...the vestibular sacs respond to the force of gravity and inform the brain about the head's orientation. The semi-circular canals respond to angular acceleration-changes in the rotation of the head-but not to steady rotation. They also respond (but rather weakly) to changes in position or to linear acceleration.*

The semicircular canals respond to rotations, the otolithio organs sense linear accelerations, including the constant linear acceleration of gravity. Both systems function using mathematical mechanics. Wikipedia says that the semicircular canals work based on the following equation:

$$\emptyset(S) = \frac{\propto S}{(T1S + 1)\,(T2S + 1)}\, q\,(S)$$

Θ ---- deflection of capula

q ---- head velocity

S ---- frequency in seconds

While discussing the vestibular system of fish, Kasumyan (2004) describes the functioning of the otolith as follows:

> *The pattern and intensity of the electric response of the hair cells is determined by the direction and value of stereocilia shifts. Electrophysiological studies have shown that the Maximum electric response arises in the group of macula cells, with the polarization axis maximally coinciding with the otolith shift direction. The response strength of the cell (amplitude of the receptor strength) is determined by the cosine of the angel between the direction of the cell polarization and the vector of the otolith shift, i.e., with an increase of the angle, the response strength drops, reaching zero at an angle of 90 (Flockm 1971; Hudspeth, 1985; Lewin et al., 1985).* (p. 5239)

I think it is very interesting that we have three semicircular canals designed to make computations in a three-dimensional world and that these canals are at right angles to each other. Doesn't sound coincidental! These systems as described have extraordinary capabilities, and why then do they seem so limited in capacity to respond to certain psychological stimuli? In other words, are there rules and laws pertaining to vestibular function that determine performance capacity and is there hope for us?

To answer these questions, let's continue our exploration of the scientific literature. Kasumyan (2004) compared growth and species' biological order in regards to vestibular size and efficiency:

> *Otoliths grow in fish throughout their entire lives, along with linear growth of the body. This distinguishes fish from other vertebrate species, for example, mammals, in which the growth of otoliths terminates before birth. (Lychakov, 1998). The size of the otoliths in conspecific fish of similar lengths may differ because the sizes of otoliths are determined by the growth and not only by the body length: otoliths are larger in slower-growing fish than in similarly sized fast-growing fish...the growth of fish involves not only increasing the size of the otoliths, but also other elements of the labyrinth: Semicircular canals, otolith cameras and Maculas, and increasing the number of sensory cells...with growth of the fish, the branches of the auditory nerve coming to the Maculas become thicker, which certainly increases the sensitivity of the organ (Sento and Furukawa 1987).* (pp. 251, 258).

So, now we see that there are different patterns of growth of the vestibular in fish as opposed to mammals. Growth rate is important. It is indicated that size of vestibular and its sensitivity correlated. We also know that the larger fish are much faster than the smaller ones. We need to compare the rules of growth of the vestibular in mammals to those of fish a bit farther. Spoor et al. (2007) had this alarming information: "Canal dimension increase with body size but with strong negative allometry i.e. the relative size of the canals tends to decrease as body mass increase." In this study, semicircular canal size was equated with agility; that is, speed of movement. This result was for the primates and all other mammals.

I am inferring that with a greater computational power in vestibular system comes a greater range of behavior options. In these studies, greater sensitivity of the vestibular is equated with greater abilities for motion. It was stated that the otoliths in primates and other mammals stop growth at birth and that the semicircular system in primates and other mammals has a negative allometry, and this was the cause of slow movement in primates and other mammals that were slow. Strangely, the Bible has a lot to say about human motion, but I have never heard a sermon on it nor seen it on any Bible study program. Our sciences are disjointed or operate in separate compartments, and religion is not considered as part of the scientific endeavor; and in fact, they are antagonistic and in competition for ownership of truth. Science is in a never-ending search for greater, more accurate truths, and religion considers itself as already in possession of all truth. These conditions have worked together and hindered both scientist and theologians and human progress. I have often wondered how man traveled so far backward that the light of the distant world seems to blind us.

In this regards the theologian Finis Jennings Dake (1999) made the following comment:

> *Perfectly Healthy Nation (Ps. 105.37)*
>
> *The reason there was not one feeble person in all their tribes was because God healed them at Marah (Ex.15:26; Ps. 107:20). Such is the will of God for men under the New Covenant, which is based upon better promises (Heb.8:6) and is more glorious in power and provisions than the old Mosaic covenant (2 cor.3:6-15). It is to the shame of the church to fall so far behind Israel in receiving promised benefits (Mt. 7:7-11; 17:20; 18:28; 21:22: Mk 9:23; 11:22-24)...*
>
> *Is it not logical that new covenant ministers and saints should meet the physical and spiritual qualifications of the old covenant (Lev.21:16-24; Dt.28)? Is the old covenant the better of the two? Are its provision and promises better? If so, then we should go back under the old covenant. What is wrong with the church that it does not get the superior blessings of the New Covenant?*
>
> *Could it not be:*
>
> *1 Because of unbelief (Mt. 17:20; Heb. 11:6; James 1:5-8)?*
>
> *2 Because of the widespread rebellion against such teaching (Tim. 4:1-8; 2 Tim. 3:5; 4:1-4)?*
>
> *Surely, it is not because God has lost His power or does not care to live up to His obligations made in the New Covenant.* (P.1053)

Here we have a theologian making a serious comparison between the health status of the church and that of the ancients Hebrews from whom the church sprang. He doesn't mention anything about a branch of healing knowledge nor does he even seem aware that there was one. It seems that somehow, the church got disconnected from the models and methods of conveying information used by Old Testament and early Church writers. Maier (1999), quoting from the writings of Philo, described how early Christians studied the Old Testament:

In each house there is a sacred chamber called "a sanctuary or monastery," where they celebrate in private the mysteries of the sacred life, bringing in nothing but law and inspired oracles of prophets, hymns, and whatever augments knowledge and true religion ... The whole period from dawn to dusk is dedicated to spiritual discipline. They read the sacred scriptures and interpret their ancestral philosophy allegorically, since they regard the literal sense as symbolic of a reality hidden beneath the surface. They also have writings by founders of their sect that they use as models of the allegorical method.

I have never taken a correspondence Bible course or studied with any church group using such methods of any such understanding as prerequisite. With a lifetime of Bible study (traditional) aided by undergraduate and graduate degrees in psychology/general science, I finally discovered numerous coded messages in the Old and New Testaments. In the Old Testament, the codes are based on the story number and the spatial position of number and motion of people, animals, and objects.

After I developed my movement program, one day, I was able to see it coded in the Old Testament in Numbers 10:11-13.

I am going to give a brief sketch such to prove that Philo was correct about the models for studying the Bible:

Phi = 1.618 = golden section and prescribes the distance to be used in the motion program.
Tribes were divided for this journey – a process of division
Twentieth day = 20 = fraction of month
Second month = 2 = fraction of year
A fraction consisting of a year and a month
These two numbers are side by side in the text = 22
Second year = 30 days + 20 days = 50 days = a fraction of the year
Everything in this story is moving. So, the .50 moves to .56 because this program is a precise biological mathematical system and .56 gives the precise number when divided The complete fraction is now 22.56
A year is 365 days 365/22.56 = 16.1790
The fraction is in motion, so move the decimal one place to the left and round the number, and you get 1.618 the golden section.

The Hebrews had an advanced health system, and part of it was a movement program. In addition, the ancient Hebrews also knew about the vestibular system and measures to take to protect it. Psalms 139 is a code description of the vestibular system and a warning that it is to be protected in the womb. I will have more to say about this later. Dake's commentary and a working knowledge of this system makes it plain that what is called the "reins" in the Bible is the vestibular. Even though we don't know why the church got disconnected from its science by abandoning its models of studying the Bible, we can see the results. What were the results of this motion science? The Bible mentions unusual abilities of agility in man on several occasions. Did these Biblical people have larger semicircular canals? In addition, modern researcher has suggested this is the case as Yang el. (2007) says, "more highly maneuverable animals do have somewhat larger canals than slower—moving animals of similar size."

Let us take a look at the Bible's reference to the agility of men:

> *For by thee I have run through a troop; and by my God have I leaped over a wall.* (Ps.18:29)
>
> *Saul and Jonathan were lovely and pleasant in their lives, and in their deaths, they were not divided: they were swifter than eagles they were stronger than lions.* (2 Samuel 1:23)
>
> *And there were three sons of Zeruiah there, Joab, and Abishai and Asahel: and Asahel was as light of foot as a wild roe.* (2 Samuel 18)
> [Dake's commentary, "as swift as a gazelle."]
>
> *And it came to pass at the seventh time, that he said, Behold, there ariseth a little cloud out of the sea, like a man's hand. And he said, go up. say unto Ahab, prepare thy chariot and get thee down, that the rain, stop thee not. And it came to pass, in the meanwhile, that heaven was black with clouds and winds and there was a great rain. And Ahab rode, and went to Jezreel. And the hand of the Lord was upon Elijah: and he girded up his loins and ran before Ahab to the entrance of Jezreel.* (1 Kings 18:44-46)
> [Dake's commentary says this was a 30-mile footrace.]
>
> *Then said Ahimaaz the son of Zadok, let me now run, and bear the king tidings, how the Lord hath avenged him of his enemies. And Joab said unto him, thou shalt not bear tidings this day, but thou salt bear tidings another day: but this day thou shalt bear no tidings, because the King's son is dead. Then said Joab to Cushi, go tell the King what thou hast seen. And Cushi bowed himself unto Joab, and ran. Then said Ahimaaz the son of Zadok yet again to Joab. But howsoever, let me, I pray thee, also run after Cushi. And Joab said, where fore wilt thou run, my son, seeing that thou hast no tidings ready? But howsoever, said he, let me run. And he said unto him Run. Then, Ahimaaz ran by the way of the plain, and over ran Cushi.* (2 Samuel 18:19-23)

The fastest horses run 40 miles per hour, and here we have Elijah out running fast horses for 30 miles without any rest breaks. Messengers ran on foot instead of horseback. These stories are confirmation that they knew about the vestibular and are proof that they had a very effective science of motion. King David, in addition to being extremely agile, was also considered a top scholar, musician, soldier, athlete, and wise (1 Samuel 16:18, 23, 17:34-36). According to the Bible, all the science and medicine reached their climax under the scholarship of King Solomon (read Dake's commentary on 1 Kings 4:29-34).

As I stated earlier, it seems that the light of science was much brighter in what we call the ancient world. They know that it took the entire solar system to create mind, and they knew about the vestibular system which modern science is validating. What are the implications for modern man?

A New Classification System

Modern Man is in the habit of observing nature, including man in their current habitat or environment, and assuming that it is there natural state of being. In other words, they think that they were always that way and that they will continue to be that way. In this regard, scientists study human beings from birth to death and note what they observed and call this the human life cycle.

Even doctors treat symptoms and sometimes what they deem to be the cause of diseases without ever giving thought to causes that could be otherwise or that a human maybe more complicated.

Modern scientists have a classification system for all living organisms. The following is an example of this system for human beings from the New York Public Library: Science Desk Reference (1995):

Taxonomic Level	Name	Example Feature
Kingdom	Animalia	Animal
Phylum	Chordata	Spinal cord
Subphylum	Vertebrata	Segmented backbone
Superclass	Tetrapoda	Four limbs
Class	Mammalia	Suckle young
Subclass	Theria	Live births
Infraclass	Eutheria	Placenta
Order	Primates	Most highly developed
Superfamily	Hominoidea	Humanlike
Family	Hominidae	Two-legged
Genus	Homo	Human
Species	Sapiens	Modern Human (p.95)

This system of classification was developed by Carolus Linnaeus in 1735 and is still used today. This is a model for the study of man and understanding of man. In order for progress to be made, this model needs to be changed and updated with additional information.

I would like to make the following suggestions allowing for future suggestions for continuous improvements. That the human classification system be amended with this box:

Human
Hybrid-reptile Vertical posture Intermittent temperature control Bipedal locomotion Verbal consciousness Different growth rates Different agility potential Non-fur bearing

Referring once again to the Jacksonian March, patients fall in two stages at first sequentially and then a sudden thrust to the ground. I believe that this is a capsule picture in reverse of the human condition. Humans at first were horizonal creatures and suddenly in a giant advance became nearly vertical walking bipeds and sequentially over time became completely vertical. Savage (1989) described a large group of early animals in which he gave the name mammal like reptiles (pp. 119-121).

Even though he believed that they were the ancestors of mammals, they remained the same and never became bipedal. I disagree, because I have an extended classification model. In addition, I would like to add that man never shed his fur; he never had any. Having come this far, are there any solutions to the problems endemic to the vestibular system? Even though, the vestibular system in humans continuously reverts back to the autistic spectrum when confronted with threats thus causing mental illness, even though it is the most powerful system in the animal kingdom. Research suggests that a large vestibular has more computing power and imparts greater agility. One is tempted to think that larger vestibular in humans along with greater agility, or movement potential would also bring a higher quality of life. The Bible suggests that humans can be much more agile than modern humans are. Research also tells us that fish are the only vertebrate whose otoliths grow their entire life; in mammals, otoliths stop growth at birth. Semicircular canals' sizes are connected to agility in that the larger they are the more agile is the animal. However, mammals and primates' canal size have negative allometry.

Man as Superhuman

I have redeveloped that ancient theory of mind. It appears that our entire solar system ending with man exists for the single purpose of mind. In fact, if you leave out the moving planets, sunlight, 1-G gravity, man's modality of motion, and man's entire nervous system, there is no mind! The solar system and motion is mind.

So, why go to all this trouble to create mind and leave mind in a state of vulnerability to malfunction causing misery and unlimited destruction? Was this intentional, or are we missing some very important information? Prominent modern evolution scientist Jones (1993) after examining the various strategies that the human body has exhibited when faced with environmental threats came to the conclusion that humans are at the end of their evolutionary road (p. 250) and that mechanisms of evolution, such as natural selection, will never make humans into super humans (p. 245). As I mentioned earlier, the problem I have with the observations are that scientists are looking at humans under what they call the human life cycle, not considering that the cycle could be changed or that humans might respond differently under different environmental conditions. The other problem is caused by classifying man as a mammalian primate with physiological restrictions involved with the vestibular growth.

Indeed, if man is as seen in the eyes of science, there is no hope. But what if the Bible's opinion of man is correct, even though Christians of today pay it no attention! In addition to mentioning the feats of agility and endurance of Old Testament Bible characters and that of Jesus in cleansing the temple, the New Testament talks about a body that couldn't be harmed and a body of glory. In Luke (10:19 and Mark 16:18), the disciples of Jesus couldn't be harmed nor poisoned. In Chapter 9:26 of Luke, Jesus mentions three different glories in verses 28-36, three of the apostles are allowed to see Jesus' body change appearance, and Moses and Elijah show up on the scene; bodies of glory. The resurrection is an event when bodies change from vile to glory. Our question is, can a human today exist in a bodily state of safety from diseases of body and mind? Can our vestibular system function at a level where it can produce safe and effective strategies when faced with difficulty? Apparently, even the Hebrews no longer possessed the knowledge of superhuman performance as Nicodemus was shocked at what Jesus did, and Jesus told him that as a leader in Israel, he should know about it.

Fortunately, there are biologists today who study biological plasticity at least in the lower animal kingdoms. Brilliant men of science are making great strides in furthering the understanding of the true nature of living organisms and evolution. The sciences of epigenetic and epigenesis have shed light on the biological change of various species due to exposure to different environmental conditions and to different conditions at different times.

My work or struggle is not to convince science that organisms are plastic, but to convince them that living humans are plastic and under what conditions. That the extinction or near extinction of the human race can be avoided by humans becoming superhumans. These scientists already know that living organism can make changes that are not random. I hope this will be helpful.

Van DeVizer et al. (2001) defines epigenetics as "the study of phenomia that led to changes in gene function that are mitotically and or meiotically transmissible without entailing a change in DNA sequence" (p.2). They further state:

> *Here, epigenetics is concerned with the broader study of the determinants of individual development as conditional, inductive interactions among the organism's constituent components and structures and between these external forces. These determinants are both temporal and spatial related to various organizational layers. As a consequence, epigenetics incorporates a developmental and an evolutionary approach as legitimately as a genetic approach.*

Modern science calls these changes that living organisms make "natural genetic engineering," and we are getting closer to answering our questions and to solutions.

As we move forward, let's take a look at words of wisdom from Shapiro (2002). Pay close attention:

> *Finally, we are beginning to obtain direct evidence for the participation of MGEs in the evolution of regulatory regions and protein coding sequences. A particularly instructive example is the sequence encoding a rodention channel (Fig.3). More than half this coding sequence derives from rodent-specific SINEs, making it a sequence that could only have evolved in rodents and not in other kinds of mammals. From the foregoing, it is evident that the capacity of living cells to carry out massive, nonrandom, genome-wide DNA rearrangements has to be incorporated into any theory of evolutionary change. If we pause to reflect that every existing organism is a survivor of an evolutionary process involving multiple possibilities of extinction then the power of natural genetic engineering should not surprise us. Organisms that can create useful genomic novelty most rapidly and effectively will have the best chance of surviving an evolutionary crisis. Indeed, the hardest proposition to accept is the assertion that organisms have not optimized their ability to expand and rewrite the information stored in their genomes. A species that depends exclusively on independent random changes for inherited novelty will not be very competitive in the evolutionary sweepstakes.* (pp. 123-124)

In the above, we are given good news and potentially bad news. Modern science recognizes the living organism can and does make physiological changes in form and function. But, the premise of this book is that the United States of America is facing a great crisis in that mental aversions are about to render the population back to its beginning status. Mental and physical health problems are going to remove the work force and overwhelm the economy.

Social conditions and family life will become too frightening to contemplate. We are facing total chaos, mass panic, and shut down. We have established that the entire human body is a mind within the constraints of the solar system, which is part of the mind. We have also established that the vestibular system as a body under restraints of what is called the human life cycle will continue to resort to the autistic end of the spectrum with its elements of mental illness and that under these conditions the human body acts as a restraint to the entire system. I am going to propose at this time that the human body can function differently under a different life cycle and facilitate viable responses by the vestibular system. A life cycle that would, in effect, make us superhuman. Let us first look at the lower animal kingdoms and the various forms of metamorphosis.

Change in Lower Living Organism

Adam, we are told, was given several commandments, and one was to name the animals, which is the same as study them. Apparently, there is much important information to be gained from such enterprise. What have modern biologists learned about animals, and in particular, what have they learned about lower species and the type of change that they undergo and the conditions that are part of the process? We are going to see that what has been observed is very instructive for human survival.

Remember that modern biologists have discovered that living organisms have a natural genetic engineering system. Keep this in mind as we continue our discussion.

Biologists talk about what is known as developmental plasticity. "This is an organism ability to inherit a range of phenotypic potentials from which the environment elicits a particular one" (Gilbert 2002, p.206), Gilbert quoting Schualhausen. Biologists observe that living organisms inherit the ability to develop differently under different environmental conditions. In some species, sex is determined by where the larva is placed and in others by variation in temperature.

In ants and bees, production of worker and reproductive caste being controlled by the hive cell placement and dietary differences. According to Gilbert (2002):

> *Developmental plasticity became a topic of great interest to evolutionary biologist. Context-dependent sex determination was seen in turtles, lizards, and fish; nutritional polyphenism were identified in ants, wasps. and moths; and predator-induced polyphenism were identified not only in invertebrates but also in vertebrates.* (p. 208)

We are aware of the butterfly and frog and their life developmental cycles in which not only does body form change, but also modes of locomotion. There are very important and highly instructive books on evolution; the proof of and the how. Here we have proof of a living organism in one form actually changing form and moving out of the water in a different form now living on land. Here is an evolutionary process taking place continually in plain sight. If that wasn't enough, we have proof that land animals changed form and took to the air; the moth and the butterfly right in plain sight. This is evolution not theory.

Once again, this is what God told Adam to do, study the animals or name them: biology! This book of life, no one can argue with. I am continuing to stay with observable phenomenon, just as I started out observing the seizure known in medicine as the Jacksonian March. It is amazing what we can know and learn by direct observation: No special technology needed.

So, the question now is, how does man change, and to what in order to solve the vestibular problem of responding on the lower end of the autistic spectrum: man's original mind? Scientists such as Keller (2002) refer to this biological problem, saying, "the basic fact is that biological mechanism is never constructed from scratch, but rather, they make use of whatever parts happen to be around-parts (or subprocesses) that had been designed to solve entirely other kinds of problems" (p.191).

Now, I have stated the problem in both common language and its scientific equivalent. A great hindrance in conceptualizing and solving this problem is that biologist have classified man as a mammalian, primate. Under this classification, man would be at an evolutionary dead end with no way out. I am certain that if we could transplant a human nervous system into a monkey or great ape that they would still be monkeys and great apes. No, not even a human brain and its vestibular system would create mind in the bodies of monkeys and apes!

Chapter 5

MOTION AND DISEASE

Looking at the scriptures that suggested at one point in history in one country everyone was healthy and people possessed superhuman athletic abilities causes one to pause and ask if the two circumstances are linked. Of course, in our modern societies, our amateur and professional athletes come nowhere near such reported feats and, as far as we know, neither does our military personnel. Among our elderly, the major cause of death is falling, which is caused by sarcopenia, the loss of muscle mass and strength leading to a loss of balance.

Everyone, including our top athletes, begin to slow down and become less agile by their mid-twenties in many cases along with a decline in their health status. Some in our society experience extreme forms of motion malfunction accompanied with mental disorders.

Modern humans are definitely not superhumans in regards to motion when compared to the characters of the Old Testament. Strangely, while the Hebrews were experiencing vigor the people of India and Greece were not. According to Bradshaw et al. (1998), the Sanskrit texts describes Parkinson's disease 3,000 years ago, and the Greek Physician Galen 2,000 years ago, and in 1817 by James Parkinson (p.307). So, these movement disorders existed and have persisted. As time has passed, more of them have plagued humanity. Bradshaw further explain this situation:

> *Victims of Parkinson disease have a forward bend of the trunk and lessened muscular power...they lack force production causing a decrease in stride length...they also have what is called a "Parkinsonian Personality"; rigid, cautious, inflexible frugal, orderly, obsessive, and depressed. Dementia also occurs later in life...*
>
> *Huntington's disease was described more than one hundred years ago by George Huntington...this disease is believed to have developed in 17th-Century Europe and then spreaded worldwide...it is characterized by unwanted semi-purposeful movement. Movements are dance like...Psychiatric symptoms include: irritability, suspiciousness, irrationality, impulsively, antisocial behavior and delusional thought...the end result is death.*
>
> *Tourette's syndrome was also described just over one hundred years ago. Movements are said to be unwanted excess hyperkinetic movements...two of the tics are hair pulling and nail biting...it is believed that autism is part of Tourette syndrome.* (pp. 307-312)

Now, we add to loss of agility and sarcopenia, Tourette, Huntington's, and Parkinson's disease. Tourette strikes during youth, and Huntington's is deadly. As can be seen it would be of great benefit for a modem science of motion that is concerned with the vestibular system in a comprehensive manner. Also, these are just additional examples of the connection of motion to mind. Overwhelming proof that motion is mind and health. Lack of motion equals lack of mind and lack of nervous system.

Schizophrenia

Mental health professionals including mind researchers regard this disorder as too complex to understand. Remember, mind is a complex system, consisting of multiple level operations from the giant solar system to the microscopic microtubule and their neurotransmitters. The whole system is a motion system, one way or another. If anything goes wrong anywhere in the system, the whole system fails at the level of mind. In addition to detecting motion and controlling movement, the vestibular is responsible for maintaining structures within the nervous system that it uses to accomplish these tasks. If the vestibular is defective, then these other structures become small and ineffective, just like a muscle from lack of use. So, just like the aforementioned motion diseases, let's add to the list schizophrenia. Schizophrenic people don't move different from the rest of the population, so scientists don't perceive it as a motion disorder. But schizophrenics do have a motion problem: They can't accurately detect the speed they are moving nor the speed of objects in their visual field what scientist call subjective motion. They also suffer disorientation or spatial disorientation. The physiological aberrations that cause these motion perception problems complex with the behavioral systems that we call schizophrenia.

So, schizophrenia just like Tourette, Huntington's, and Parkinson disease is a disease of motion with its own set of behavioral abnormalities.

I have known a lot of schizophrenics, and they all have problems with directions so much so that some refuse to drive, and others will drive only in small towns. Another trait is that you don't want to give them a ride as they will insist that you are driving 80 miles per hour when you are driving only 40. This personal experience with them outside of an office or patient room gives you an additional perspective on this condition. Neuroscientist now know that schizophrenics have brain abnormalities when they initially thought there was no physical basis for this disease. It was thought just to be a consequence of something in one's life experience. This is the definition of schizophrenia given by Goldstein et al. (1980), he states:

> *This large category includes a group of disorders manifested by characteristic disturbances of thinking, mood and behavior. Disturbances in thinking are marked by alterations of concept formation which may lead to misinterpretation of reality and sometimes to delusions (fixed but erroneous beliefs) and hallucinations (false sensory experiences), which frequently appear psychologically self-protective. Corollary mood changes include ambivalent, constricted and inappropriate emotional responsiveness and loss of empathy with others. Behavior may be withdrawn regressive and bizarre. The schizophrenic in which the mental status is attributable primarily to a thought disorder, are to be distinguished from the major affective illnesses which are dominated by a mood disorder.* (pp. 33; 101)

As you can see this basic definition says nothing about schizophrenia as a motion perception disorder, and as a result, this disorder isn't put in its proper classification as one of the motion disorders and isn't understood.

Case Histories of Schizophrenia: Its Cause

I had the unfortunate experience of growing up in the house with a schizophrenic. At the age of 28, this single mother of two sons, single as a result of a separation from her husband, had been living with her mother for several years. At 28, she suddenly had a serious break with reality. A very attractive woman who all of a sudden stopped caring for her hygiene. Started talking crazy, eating only a piece of bread and bacon, and sitting alone and talking to herself. She picked at her hair and broke out laughing even though no one was talking. At night, she walked around with a hammer. She believed she was pregnant even though she wasn't. She was diagnosed as chronic schizophrenic, and she never recovered. Her mother told me about an experience that she had while pregnant with her daughter. She said that she and her husband were arguing and that her husband got a shot gun to shoot her with, and she grabbed the barrel of the gun and that her husband lifted her and turned her over and over until she fell and released the gun. She said she always believed that her daughter was ill affected in the womb and caused her to be mentally ill.

She was 100 percent correct. Overwhelming centrifugal or rotational forces subjected upon a fetus is too much for a young vestibular system computing power and can damage the system. In fact, Moyer (2014) expressed serious concern for small children as she advises, "Do your best to avoid activities with excessive spinning or twisting as movements in these planes can have negative effects including over stimulation, lethargy, or change in heart rate or breathing (p.2). If this type of motion is dangerous for small children, then imagine its effects on a fetus. We already talked about the heart rate of a fetus and its connection to basic personality of the new born. A person who has experienced overstimulation of the vestibular while in the womb won't have the personality resources nor the nervous system to face up to the life stresses of young adulthood. This woman in her twenties couldn't handle a broken marriage and becoming a single mother with a malfunctioning vestibular.

This woman had a nephew who also became schizophrenic in his twenties. This man's father purchased a truck while his wife was pregnant with him. He drove his wife to his mother-in-law home from time to time. This trip was taken on a very hilly road. In addition, the father was fun of extreme accelerations. The rapid up and down of the hills and the extreme accelerations on the straightaway was too much for the vestibular of the fetus in this case as well. This man never dated nor did he marry. He is currently living in a shelter for the mentally ill.

Does the fact that these two people share their gene pool mean anything in regards to their common ailment, or was it their common experiences of being exposed to motion as a fetus that was dangerous to their vestibular?

Etiology and Hysteria

All the way through school and college, I had no use for the study of history. However, it was in college that an incident occurred that would after spark my interest in the subject. While standing outside a history class waiting for it to finish, so I could be seated for a chemistry class, I heard part of a lecture that really got my attention. I never forgot that lecture, and I even bought the book on the subject of the lecture and read the entire book after graduate school. I have been a history buff ever since.

Yes, history is a very important subject. Now the only disappointing thing about history is that it only goes back so far before information gets sparse and even confusing. However, for the purpose of this subject enough information is available. Goldstein et el. (1980) had this to say about hysteria:

> *For reasons that are not clear to us today, many people in Eighteen and Nineteenth—Century Europe suffered from physical incapacities that make absolutely no anatomical sense, such as glove anesthesia, in which the person has no feeling in her hand but does have sensation from the wrist up into the arm... From the time of the ancient Greeks medical science believed glove anesthesia and other so-called hysterical reactions stemmed from a physical disorder. They further believed that only women had this disorder and alleged the cause to be a wandering uterus.* (p.35)

If you are a student of history in general and Europe in particular, then you know that man has been involved in a constant struggle to improve his living condition. During the time span mentioned by Goldstein, it has been Europe and America that were involved in developing and improving modes of transportation. Also during this time period as modes of transportation improved, they also became more and more accessible to the public as they went faster and further on both land and water.

During this time, all sciences were in an early stage of development, and the science of technology seems to have advanced faster than the biological sciences. Even today, the only concern of transportation technology is the prevention and survival of crashes and of course more speed. As I have said before, humans are limited. There are limits to what we can do, how much we can do, and when we can do it. During the period mentioned, the life sciences weren't advanced enough to be connected to motion technology. The only concern at that time was that transportation modes would be reliable and affordable for the general public. It could not have been otherwise. This situation has changed as the life sciences have made great strides, but more on this later.

Now that we have advanced in knowledge, there are several things we need to do. First, we have got to recognize man or mind as outlined in this book, and that mental and physical maladies are connected with the vestibular system, which is also limited. Inventors of the eighteen and nineteenth century Europe knew nothing of the vestibular system and its connection to motion and human mental and physical health. However, modern neuroscientists (at least some of them) know well that vestibular malfunction causes mental illness including, hysteria disorders, including problems with hand illusions. Smith et el. (2013) quoting Lopez, "Lopez and colleagues have reported that vestibular stimulation can modify the body schema and the sense of ownership of an illusionary hand in the "vubar hand illusion" (Lopez et al., 2010, 2012; p.4). The evidence is clear and sufficient, however, currently ignored. Secondly, all the great scientists currently researching this subject from different disciplines and perspectives need to get on the same page and develop new treatment strategies programs and preventive measures. The transportation industries need to get on board by developing vehicles and other modes of transportation that're safe from injury and accelerations. There is room for a lot of new technologies.

Finally, we have to somehow overcome the obstacles of apathy and resignation. We may be limited, but we have not exhausted the means at our disposal. Knowledge has increased, and there is no excuse for us not to use it. We don't have to sit idle waiting for the Rapture when there is much good, we can do. Did God intentionally go to all this trouble to create mind just so he has to destroy it? The people in the church are subject to and being overrun by these same physical and mental maladies in the general society. I have already touched on the comparison of the health status of the Old and New Testament churches. There are still other things that need to be done some require group effort, and others can be accomplished at the individual level. More later.

Human Movement and Disease

We have already covered the class of movement diseases that befall modern humans that are not a result of their own fault. This list was also completed by the addition of schizophrenia so that it could be properly understood. It consists of Tourette Syndrome, Huntington's disease, Parkinson's disease, and schizophrenia. Since people with schizophrenia move like everyone else, it isn't apparent that they have problems with motion and spatial disorientation that is at the root of their behavior patterns.

You guessed it; in addition to mental diseases being connected to movement disorders, there is a host of physical disorders associated with intentional human movement patterns such as jogging and speed programs. Currently, a small number of medical scientists are aware of the danger of jogging programs and other aerobic programs. This situation first caught my attention as people were falling dead while jogging and while working out for football. From my own personal experiences over the years, I had learned not to jog. However, it took a while before I would fully understand why. While I was learning from direct experience and observation, there were medical experts working directly with Olympic athletes, armed with medical knowledge and equipment. Dr. Irving Dardik is considered the medical pioneer using his principle of "Super Waves" to develop a new fitness and wellness theory and program. The following is a very important statement by Dardik (2016) on the subject of disease etiology:

> *In this regard, the implication for fitness and for space travel are significant. when we speak of fitness, we infer a positive relationship between fitness and health: fitness should cause good health. However, this is not always the case. In fact, there are reports that suggest an association between continuous long duration exercise with infectious diseases including polio, echoviral meningitis, respiratory and gastrointestinal infection, as well as with cancer, cardiovascular disease, and multiple sclerosis, 2-, 29, 30, 31.*
>
> *This association between fitness training and chronic disorders can now be explained by recognizing that prolonged continuous training in a narrow target heart rate zone, through aerobic exercise or any other form of linear stress actually will train HRV to decrease. The same would apply to any behavior done in a continuous fashion. By teaching HRV to become more linear, more even in other words, flattening the heart waves the individual is actually training to become more susceptible to chronic disease and all-cause mortality.*
>
> *Instead of linear aerobic fitness, which is designed to push the limits of oxygen consumption (Vo2 max) but ignores oxygen recovery, we should be training the heart wave range to expand. In fact, physical fitness (like energy) generally is defined as "the ability to do work." What this definition*

does not take into account is "the ability to not do work," i.e., recovery. In contrast to linear fitness, Wave Fitness TM, a new program designed to expand the heart wave range, incorporates both work and recovery as a wave. From this new wave perspective, Vo2 max is a partial linear measurement that reflects increasing oxidation (energy expenditure), which, in nature's reality, is only one "side of the wave." The other side of the wave is a decrease in oxygen consumption that reflects antioxidation (energy recovery). Therefore, in nature, oxidation and antioxidation are inherently continuous waves. The mystery of HRV is solved, and the ground work is laid for a whole new approach to preventing and reversing chronic disease as well as optimizing health performance and longevity by expanding the heart wave range (Figures 5 and 6; p.34)

Also, I would like to add that these long duration aerobic programs are antagonistic to muscle development and maintenance, which is a hindrance to the immune system and balance in the elderly.

So far, we have seen that motion is an extremely important subject for human well-being. Motion with acceleration can be dangerous to a human fetus and motion of low velocity of long duration that is repeated is dangerous to people of all ages. Once again, heart rate is seen to be a major controlling factor in regards to psychology and physiology of humans, and once again, it is the vestibular that is in control of heart rate. Other researchers have gotten involved with Dardik's work and the appropriate mathematics of this system has been delineated, and it is the same math of my movement system, which is the Goldean Mean or phi 1.618. It was while I was reading everything that I could find on the Goldean Mean while still getting greater understanding of my program that I accidentally came across Dardik and Dan Winter on the internet. Dardik's program is designed to improve the capacity of the heart and lungs while mine is designed to improve muscle mass and agility (footspeed) and to maintain these qualities throughout the lifespan. My program also uses heart rate consideration and is much more complex.

Nevertheless, there is still a lot of work left to be done. More on this in the last chapter.

Chapter 6

MIND AT THE LEVEL OF BRAIN

So far, I have built up a case for the philosophy that brain is not equated with mind, that mind is a multilevel, complex entity. However, that being proved, mind cannot exist without any level out of sync or not functioning properly for whatever reason. First and foremost, the brain is dependent on stimulation from the vestibular for its form and function.

It's chemical flow and neurotransmitters are first under the control of the circadian cycle as stated, "When humans are exposed to a daily dark/light cycle, cyclic production of specific neurohormones and neuropeptides result in changes in sleep and alertness, body temperature and pressure, metabolism, and reproduction" (Roberts, 2015 P.2).

This requires that we be in harmony with the motion of our solar system and that we have the proper amount of protein in our bodies. Lehninger (1975), had this to say:

> *Although the brain is not capable of regeneration and its cells do not turn over there is constant biosynthesis of certain brain proteins from amino acids, particularly the protein subunits of microtubules (page 771) and microfilaments (page 771), which together constitute more than half of the soluble proteins of developing brain and participate in axoplasmic flow. Energy is also required for the synthesis of acetylcholine and other neurotransmitter substances. The brain uses energy at a constant rate; active thought or intense concentration cause no significant increase in oxygen consumption.* (p. 838)

In addition to protein being important for the development of and maintenance of muscle mass, which is important for the immune system and movement and balance in the elderly. It is critical for brain structure and function. It is critical that, there is enough amino acids available for all the things the body does and still have enough for the brain. The body has to make many molecules from amino acids other than those used for brain maintenance, neurotransmitters and muscle mass.

Bern (1990) gave this information:

> *In addition to their incorporation into proteins, many of the amino acids, including some of the essential ones, are precursors for important molecules, such as purines, pyrimidines, polyamines, phospholipids, creatine, carnitine, methyl donors, thyroid and catecholamine hormones, and neurotransmitters.* (p.498)

Our brains being one of the levels of mind, but not mind itself, is a necessary component of mind and has critical protein needs for structures and chemicals and the production of chemicals under the control of the circadian system and important structures under the stimulus of the vestibular system.

Brain Structure and Mind

As mind research continues to advance, one common denominator in all these conditions: Cushing syndrome, depression, post-traumatic stress disorder, Alzheimer's disease, schizophrenia, and dementia all co-exist with hippocampal atrophy. In Canlon (1999), a chapter is devoted to this subject, and Awad (2007), a whole book. According to Brandt et el. (2005), "MRI volumetric revealed a 16.9% decrease in total hippocampal volume in BVL patients relative to controls…" (p.2734). Patients with bilateral vestibular deficits, have trouble with locomotion. These were adults who acquired chronic bilateral vestibular loss. We see that vestibular dysfunction affects adults by causing motor locomotion problems as well as mental problems. In regards to children, Wiener-Vacher et.al (2013) states, "Bilateral loss of vestibular function at or close to birth results in motor developmental delays."

I believe that Wiener-Vacher et al. (2013) do an excellent job explaining the relationship between the vestibular system, hippocampus and mind:

> *Vestibular inputs provide several types of information which are respectively engaged for diverse corresponding sensorimotor and cognitive processing (Potegal, 1982; Weiner and Berthoz, 1993; Borel et al., 2008). The three pairs of semicircular canals and the otolith organs provide complementary information about several types of rotational and translational head movements involving accelerations, and importantly, signal the direction of gravitational force. The latter is fundamental to terrestrial life from birth since postural adjustments and active movements invariably must take gravity into account. The brain builds representations of verticality based upon vestibular somatosensory proprioceptive, and visual information, constructing a "geocentric reference frame (Borel et al., 2008). Vestibular patients are thus handicapped in acquiring information during active displacements in the environment since sensory frames of reference (e.g., visualor proprioceptive; Lacour et al., 1997; Isableu et al., 2010) must be established without vestibular information. Furthermore, gravitational and other vestibular information can be instrumental in timely acquisition of fundamental spatial relationships of up vs. down, left vs. right, front vs. back, etc. Wiener - Vacher et al, 20122). Infants first learn these spatial relations with reference to their own body. Understanding spatial relations between the*

body parts is difficult for vestibular impaired infants, perhaps in part because proprioceptive information about gravity is not reinforced by otolithic gravity sensation.

This would then have a negative impact on understanding other spatial relationships such as over/under, inside/outside, within/without interposition, etc., these patients would then have difficulty applying these concepts for establishing coordinate, systems for representing the relative positions and orientation between environmental features and their own relative position and orientation to all of this. Furthermore, if concepts like:

...close + distant, superior, inferior, etc., are poorly understood, the child could also have difficulty extending them to arithmetic and geometry as well as to other non-spatial domains where sets and groups are compared (e.q.) syntax, biology, history ... One theoretical framework of how cognitive representations emerge in the brain contends that sensorimotor loop activity is internally simulated and represented in the absence of the relevant sensory inputs and movement. This would lead to anticipatory processes and the construction of yet higher levels of representations.' Since vestibular dysfunction would impair the many sensorimotor processes described in the previous paragraphs, serious consequences can be expected in building representations and cognitive processing in the corresponding functional domains. (N.P.)

I would like to add that my life experiences in different work settings seem to verify this theory. While working with eight- to 12-year-olds at a psychiatric facility, I found that none of the children could do a push up. They didn't know how to position or move their body parts in order to get in the push up position. An 18-year-old retarded person in jail had the same problem. While working with a group of college football players on footspeed, I found that those doing poorly in college also had problems feeling their arm position and controlling them. Mind, once we get to the biological level, is a multilevel function.

The brain and the rest of the nervous system seem to be tools that the vestibular system uses to create mind at the biological level. If you have problems navigating in space, which is the foundational level of mind in living organisms, then you are missing mind foundational building block.

A long time ago, a gigantic motion system was put in place to produce what is known as mind. Computing motion by the vestibular system aided by the circuitry of the hippocampal structure of sufficient size and volume in a human form creates mind. The vestibular system appears to have three critical stages of development in humans. These stages are inferred from the fact that hippocampal size is dependent upon vestibular stimulation. Wiener-Vacher etal., (2013) say in regards to hippocampal development in children, "The data presented above suggests that different types of growth and maturation are occurring in the period up to age 2-3, then leading up to 6-8. and then up to the age of 11 where adult-like characteristics appear" (N.P.). This is significant and will be discussed farther in chapter seven.

Chapter 7

THE NEXT STEP: EXPLOITING HUMAN PHENOTYPIC PLASTICITY

"Man know thyself," or, "What is man that thou art mindful of him?" Two very important statements in regards to our subject. In an effort to know about life and living organism, scientists have studied every life form possible and are still in the process. They make great efforts to study and understand every aspect of every creature's existence. Of course, all of this is done within the context of what is called the natural environment. In some cases, various species alter environmental conditions, and scientists observe the responses or adaptations that these particular organisms make. Even from this limited process scientist have observed a large range of responses among living organisms. So far in the human species the differences observed are: race, size, health status, IQ, athleticism, and life span. Scientists consider man's fate in regards to the stated differences as being determined and fixed by their genes inherited before birth and that activities such as bodybuilding and weightlifting are under the restraints of genetic variability. In this regards, health status and length of life are also seen as components of genetic expressions.

I am going to ask the question: Are human genes limited or handcuffed to what we observe? Can human beings respond to environmental variations in ways not currently known? If so, how can the body respond and what conditions do they take place? In this chapter, I intend to answer these concerns. There are hindrances in our society against advancing this work. Only certain people have access to an outstanding education for various reasons. And as more and more people are born autistic or with other developmental problems, the brain pool is getting smaller, and more and more resources are being used just to deal with the problems. Christians are occupied waiting on the Rapture and much of science and the public, view our current situation as normal: birth, aging, disease, and death. In addition, any attempts at human engineering seems to be looked at with fright especially since the German atrocities of WWII. However, my work isn't intended for propaganda for the advancement of one group at the expense of others, but for the advancement of the human race in the process, avoiding the extinction of the human race in addition to the attainment of much healthier, longer lives, and hopefully peace and safety for all.

Man's Biblical Responsibilities

I would like to say here that this book isn't intended to be an exhaustive work on this subject. Hopefully, it will start the process of solutions for our present problems.

We have got to stop producing more and more angry, selfish, and violent people. This process starts during pregnancy and is completed in the education system. What does the Bible commands us to do?

I am touching on this subject because I keep running into Christians who believe that all we can do is prepare for the Rapture because only Jesus can solve these problems. I can't afford to waste time over various interpretations; time is too valuable. However, I would like to remind my Christian friends that Jesus Himself promised His disciples that when they could understand He would send a teacher who would reveal all truth (John 16:12-5). Then again, Jesus prayed that His disciples not be taken out of the world, but that they be protected from evil (John 17:15). Our Bible does give guidelines to protect us while living on Earth.

First, the Bible gives two special instructions for pregnant women that we don't follow today that would facilitate healthy births with functional vestibular systems. I reported earlier in this book in Psalms 139 chapter, and in particular verse 13 and Dake's Commentary, verse 13 say, "For thou hast possessed my reins: thou hast covered me in my mothers' womb." Dake (2006) commentary in column U reads:

> *The Hebrews believed that the reins were the first part of the human fetus to be formed, so here it may refer to laying the foundation of being. This seems clear from the fact that it speaks of the formation in the womb (v.13) and of being fearfully and wonderfully made* (v.14). (p.1018)

In our modern language and science, the reins mentioned here are nothing but the vestibular system. The psalmist here mentions the protection that was given while a developing fetus. In modern societies, we make some effort to protect the unborn from what is known as teratogenic agents. Hallahan etc. (1999) defined these as "agents that cause abnormal growth or malformation in the fetus" (p.54).

They discussed three agents: alcohol, cocaine, and lead. The Bible does include alcohol as a teratogenic agent, but also a certain class of proteins. The Bible is very specific about teratogenic agents, which include all proteins not on the clean protein list and alcohol (Judges Chapter 13:3-4), but we would have to include protection against forces that damages the vestibular specifically. Ezekiel Chapter 13 is a command to design and build houses that are safe during violent weather systems. Houses should be safe against structural damages, as well as protect adults and fetuses from changes in atmospheres pressure. The United States has a peculiar weather system. As I watched a program on the weather channel, it was stated that the US has more tornadoes than the rest of the world combined. Thunderstorm systems are a frequent occurrence in America subjecting its population to numerous changes in atmospheric pressure. I would include this situation as a threat to

National Security. According to Atwater (1999), human activity of our era is contributing to the weather problem:

> *Measurements from remote sensing satellites and field observation confirm that fluctuation in cattle grazing patterns can alter local climate to a degree that exceed the effects of greenhouse warming or urbanization. Studies of various climatology records—when compared to a given area, the people there and historical or current events—suggest a possible link between weather fluctuating patterns on attitude and behavior from local residents.* (p.189)

This sounds like we are making a very bad situation even worst with our modern way of life. The non-native people from Europe and Africa who first came to settle in the New World knew nothing about this country and possessed no information to do anything about it if they had. We live in America with no regard to the nature of man nor to the nature of our environment and its impact on behavioral health and health in general.

The Bible gives incidents of change in local weather causing change in human behavior. In Exodus 20:18, King James Version we have this statement:

> *18. And the people saw the thundering and the lightings and the noise of the trumpet, and the mountain smoking; and when the people saw it they removed, and stood afar off.*

Here we have a case of sensory mixing people could see the sound of the trumpets. There are other incidents mentioned, however, I think this one is enough.

So, the Bible Command us to:

- Protect the vestibular system while the fetus is developing in the womb.
- To protect the vestibular of fetuses from what it calls unclean animal proteins
- To build housing that protects humans from changes in atmospheric pressure which can damage vestibular organs.

If you have ever flown in a plane then you know about the effect of changing altitude pressure has on your ears. In my case the pain is unbearable. One of the purposes of mind is to participate in our own salvation. We can make life on planet Earth much more pleasant.

Additional Dangers to the Vestibular

As we move to take action to protect our vestibular organs, and in so doing our species and the full potential of what it means to be human, we have to be aware of all the dangers to the vestibular and do what is necessary to prevent them. We have already mentioned the dangers of exposing a fetus to accelerations of any kind as well as children.

We have to remember that the vestibular is an extremely sensitive system. On one hand, this great sensitivity facilitates mind; on the other, it is very vulnerable to dysfunction: it can be overwhelmed and damaged when exposed to stimuli that it isn't designed to process or hasn't been programmed to function.

I can remember growing up as a youngster I would hear old black people complaining after eating pork. They complain of their blood pressure rising, headaches, dizziness, and my grandmother said the room was going around and around with her. All of these symptoms would sound like vestibular dysfunction. The commandments or dietary laws laid out in (Lev.11:1-30, Dt. 14:3-20) must be observed for good mental physical health, and as we shall see later especially for black people. These old black people didn't have medical degrees and equipment, and neither did I, but we all had enough sense to relate cause and effect. In fact, medical doctors were advising them of negative effects of pork on those who were having problems with high blood pressure; however, they kept eating it and dying from heart attacks and strokes. Today's scientists call the proteins on the Biblical clean list as fast burning proteins as opposed to slow burning proteins a very important distinction.

Remember, life on Earth is limited including human beings. All of our biological or physiological functions are guided by a 24-hour clock. It stands to reason that metabolic functions need to keep pace with all of the bodies physiological processes. Proteins need to be available when they are needed. Also, is the problem that medical research has revealed that some forms of proteins and amino acid profiles can be dangerous (Moran, 2002; p.54).

In our modern society, we have vehicles that can produce large acceleration force: such as cars, trucks, boats, trains, and airplanes. They also allow us to cross time zones in relative short periods of time, which I definitely advise pregnant women to avoid as well as small children. Modern societies also have amusement parks, rooms, and fairs where machines are that produce large angular, rotational, forces and roller coasters that produce large up and down as well as linear and rotational accelerations which are very dangerous and to be avoided by pregnant women and small children. There is a desperate need for the transportation industries to develop technologies that can control forces inside, trucks, cars, boats, trains, and planes. Until then, individuals need to take individual action as to how they drive to avoid rapid accelerations as much as possible. Pregnant women need to take every precaution to prevent forces from injuring the unborn. Research in America is going to have to discover how we need to modify our collective life behaviors to get our weather under control.

Human Physiological Diversity

I would like to re-emphasize that the purpose of this book is to enhance health for the entire American population and not to stir hate. Emphasis on the health of the entire population and the population consists of multiple racial groups. Black Americans more than any other race seem to be suffering the most in almost every disease category. Even if we had economic parity in this nation this would still be the case. Black people, having suffered under discrimination don't want to hear about racial differences or being treated different in the medical system are being advised by medical establishment to their detriment. Much research has and is being done on this subject; however, none of it is being used to treat black people nor to develop a model for treatment.

Writers such as Barnes (1988) and King (1994) have provided significant information about black physiology. The internal melanin inside black people's bodies interact with drugs and often produce harmful effects and the melanin in the gut also interact with drugs and can prevent the intended affects in other parts of their bodies. They also showed that the black physiology requires a different diet other than those espoused by our medical community. I would like to add that since you have melanin in mitochondria of blacks, that this definitely indicates greater metabolic capacity, and this is why I advise adult blacks to restrict their proteins to the clean protein list in the Old Testament, so they can be in what I call metabolic rhythm.

Harris (2006) in a symposium on vitamin D, and African Americans proposed that vitamin D deficiencies in black Americans were responsible for the poor health status of this population. It has also been determined that D3 is important for brain, heart, and liver function. The problem is a result of having dark skin people living at a latitude where sunlight is of insufficient intensity. White America and other fair skinned people here don't have this problem as their skin can produce more vitamin D when exposed to sunlight. So, if you are a black American, you need vitamin D3 supplementation for your body to function and for mental health. I would like to add considerably more D3 than recommended. You need to conduct research!

The Bible is full of scriptures advocating more light. Since the only people in need of exposure to a large amount of intense light and suffer deficiencies in vitamin D3, currently you have to ask who were the Hebrew prophets addressing concerning this issue. Both Isiah and Ezekiel put great emphasis on this subject. Isiah 30:25-56 talk about altitude and a sevenfold increase in light as means of healing. Of course, the higher you go up on the Earth, the greater the light intensity. Ezekiel went so far as to describe the physical appearance of man who was upon the very high mountain (altitude) who was measuring the future temple sight. Ezekiel 40:3, "And he bought me thither and, behold, there was a man, whose appearance was like the appearance of brass, with a line of flax in his hand, and a measuring reed; and he stood in the gate."

Ezekiel Chapter 40 to Chapter 48 describe this temple. The man on this mountain giving the measurements was a man of color (brass). Once again, you have altitude, and this time, a diet high in fish, the only dietary source of vitamin D. You have a man of color, high intensity sunlight, and dietary source of vitamin D. What else do I need to say except Ez.43:10-12 emphasizes pattern, form, and measurement, and this, you will see later, is very important. How do we overcome the human tendency to rush, to skip details and process, and to put things in place without knowledge? I have often wondered why we only use part of the Bible. Even the Bible teaches that health isn't possible without careful human engineering.

Engineering the Environment and Man

So far, I have tried to establish that mind is a giant complex machine. At the biological level, mind is made possible by a vestibular system operating within a human form. That the vestibular and mind, as we currently experience them, are subject to mental and physical difficulties. We looked at vestibular development in the animal world suggest that mammals and thus primates are at a dead end for vestibular enhancement. However, I showed that humans have been misclassified and are in fact, humans are hybrid reptiles. Being reptiles is actually a good thing and suggests that we can improve our vestibular system, but first we need to protect the system from accelerations from storms, modes of transportation and from recreational forms of acceleration. Also, pregnant mothers are to avoid alcohol, drugs, and slow proteins. Adult people of color also to avoid drugs and slow proteins. In addition, black American suffer vitamin D3 deficiency. We will have to take measures to deal with this just to produce what we consider normal people with normal life spans. I would consider this as a level one engineering, just for the survival of America.

In the light of modern biology, especially developmental biology, are there any means where by humans can be enhanced? Is there a level two for human development? Morange (2002) said this about modern biology:

> *...the fertilized egg does not contain a complete description of the future organism, as assumed by preformationists, but rather the coded instructions required to produce its molecular structures and to bring them into operation in time and space.* (p.55)

Remember biologists have determined that living organisms can express different sets of genes, depending on the environment or external conditions. Is this true for humans? Based on my personal experience and experiments, definitely!

I am aware that, there are scientist who do life extension research and sports scientists who research ways to improve human performance; however, their work isn't connected in any way. So far, I have been begging the question are the two related? In our day and time, people sometimes live to be over 100 years old, but none of them would be classified as world class athletes. Granted some of them are strong enough to walk and can maintain balance while doing so. These centurions are in a process of decay like the rest of the population and don't live for another century. The elderly suffers a condition known as sarcopenia—loss of muscle mass. I call this condition adverse metamorphosis. I believe that muscle mass, growth, and growth rates are keys to engineering the human body and that this may be a means of improving man in addition to enhancing the performance of our vestibular system or because it is enhanced in the process.

While training a prepubescent to play little league baseball with my program, something happened that I wasn't expecting. I already knew that footspeed could be improved drastically. However, I didn't know that you could change the form and function of an adolescent's body. I've learned a lot more since that experience, enough for me to propose the following. First, I want to give these two definitions by Wilson (1999). Metamorphosis, "A radical change in body form, physiology, and behavior during the growth and development of an organism" (p.402). Phenotype, "The observed traits of an organism created by an interaction of the organism's genotype (hereditary material) and the environment in which it developed" (p.404). I am going to propose that humans currently go through a stage of metamorphosis in that aging is accompanied by loss of muscle mass, strength, and speed. This is a change in form and function and a hindrance to the vestibular system, which is what we want to improve.

The goal of human engineering is to maintain and enhance muscle mass and agility throughout the life span and hopefully greatly lengthen life and the quality of life. Based on the material already covered in this book, our means of health, including mental health, is through the vestibular system, so our goals are to protect the vestibular and then to enhance it, and this I believe is within our power to do so. I was really shocked when I stumbled upon the work of Irving Dardik and Dan Winter. The fact that the Goldean Mean mathematics is at the heart of his and my work. Dardik is concerned with health and vitality, and I was concerned with footspeed. With my expanded understanding of the connection between the vestibular, health, agility, and body form, and observing how my program changes body form regardless of age I am hopeful with reason. Even though, I can't find anything wrong with Dardik's program and theory, my program can do what his does and considerably more. Make no mistake.

Dr. Dardik's work is extremely important. The fact that we both ran into Goldean Mean (1.618) mathematics is significant. Remember, we have already established the relationships between vestibular size and sensitivity and agility and body size.

It took me 10 years after studying the data I had collected from my own training and others before I came up with the following equations:

Systolic blood pressure = 4.5x (Resting Pulse Rate/100 Meters Time)
Diastolic blood pressure = 1.5x (Resting Pulse Rate/60 Meters Time)

I call this system the asymmetric grid. A complex training system that not only greatly increases footspeed but greatly aids muscle building. You can see from a visual inspection that it includes heart function in relation to time and distance being 100 meters and 60 meters or 1.6 or an approximation of the Goldean Mean. The equations appear simple; however, they create boundaries that ensure all other quantitative physiological functions aren't exceeded. Combined with a modified dietary program facilitating continuous and rapid lean muscle mass you have the means to control and modify human growth rates and thus body form.

Since I found my program coded in the Bible as well as other codes, I know that someone a long time ago carried out the entire program. So, I know I can trust the Bible for correct instructions. The Bible seems to suggest that a human must remain in a stage of development for one hundred years, not live to be 100 in a state of decay. (Isiah 65:20). I already know how to change the form and function of a human, and I am still experimenting at the age of 65, hoping to live long enough to reap the benefits. I have learned more than I ever dreamed I would, and there is a lot more for me to learn.

Some life extension researchers advocate a restricted calorie diet for life. However, even nature teaches us that when creatures change phenotypes, there is also a change in diet, such as in the case of the butterfly and frog. If humans could enhance human performance and maintain it throughout longevity with a single phenotype that approach might be feasible. But experience does not support this theory. So, I am suggesting a different approach by which health and longevity are to be obtained by going through a life cycle of multiple phenotypes, and the human body can change more times than the butterfly and frog with great benefit. We observe that even simple life forms changing form and function to a variety of stimuli at different times of their life cycle and with different environments. By these means, we (humans) are to avoid movement disorders that cause or co-exist with mental and physical disorders. I have had success with experiments, even though much more work needs to be done. At this time, I can recommend the following stages (natural genetic engineering system; human):

Form – human weight, height, physical and mental capacity in conjunction with growth rates
Form————————————Stage 1
Birth————————————-9 years old

Restricted calories
Vitamin and mineral supplementation
Slow growth in height
Slow growth in body mass
Emphasize development of nervous system
Head growing at a greater rate than body
Height at 9 years 4'6", weight 461bs.
At age 2 start foot speed development MOVE PUBERTY to an earlier stage
Form——————————————-Stage 2(a)
9————————————————-14 years
Change in diet and training program
Accelerated growth: increase in lean muscle mass and height
Increase strength and foot speed
Height at 14, 6' + or - 2"
Weight 235-2401bs.
Body grows at a greater rate than head
Form———————————————Stage 2(b) alternative
13———————————————14 years of age
Age 13 weight 1371bs., height 5'7"
Change in diet and training program
Dramatic change in height, body mass, strength, and foot speed within one year
Age 14 weight 235-2401bs. With a height of 5'11"
Form———————————————Stage 3
14———————————————35 years of age
Maintain size 235-240lhs.
Emphasize strength increase and foot speed
Increase vitamin and mineral supplementation and protein intake
Pay attention to joint health
Form———————————————Stage 4
Pay attention to joint health and accelerate muscle mass growth up to the age of 50 Continue to increase foot speed
Dramatically Increase intellectual activity

As you have already guessed, it would take another book to provide the details of this system. As you can see, human form is constantly undergoing change, but controlled change. We already know that pygmies have characteristic growth rates and that they have short life spans. In contrast, our giant basketball players have certain growth rate patterns, and they as a group have shorter life spans than average. So, we are controlling both form and function. The purpose of the muscle mass is to enhance immune functions, balance, agility, and the vestibular system. Remember, protect, preserve, and increase the capacity of your vestibular system. This is a complex approach to an abundant life!

More and more Americans are being born at a bad part of the autism spectrum, and they have personalities composed of great irritability and even worse, anger, which is leading to more and more hate and violence. And they are functional to various degree, and this is creating more violence and a smaller and smaller work force. Those of us who can contribute to the solution should do all within our ability to do so. Science has finally advanced far enough for us to understand more of our Bible and science itself provides irrefutable proof of a creator. It takes a whole universe for mind to exist, and the creator expects us that can to use mind in the salvation of man!

Ultimately, the mind is a moving observational system. It is moving on a planet that has a moving atmosphere and is part of a moving solar system. All physical, chemical, and biological processes in a human form are directly related to this condition. Motion in nature isn't uniform but redundant, causing mind to be subject to relativity and uncertainty principle. This allows for the mind limitation that humans currently experience and the potential for a degree of enhancement. These changes in the laws of motion in nature are guides for methods for the enhancement and advance of the human race.

Even the Bible instructed the ancient Hebrews to produce advance technology for the dual purpose of protecting and engineering their physiology. Dire consequences were predicted for the nation if this program was neglected and punishment for the priest and prophets (scientists) if they refused to guide the people and the process. These technologies were to be designed to control the effects of the Earth's atmosphere on human beings and to be used for design of multiple human phenotypes in order to enhance the body, and thus the mind. The Earth's atmosphere and climate hasn't always been as it is today, and its current state isn't a permanent affair; it will change. This is so important. Let us look at some Biblical references pertaining to it, all scriptures from the (NKJV, 2006):

> *When the Lord has washed away the filth of the daughters of Zion, and purged the blood of Jerusalem from her midst, by the spirit of judgment and by the spirit of burning. When the Lord will create above every dwelling place of Mount Zion, and above her assemblies a cloud and smoke by day and the shining of a flaming fire by night. For over all the glory there will be a covering. And there will be a tabernacle for shade in the daytime from the heat, for a place of refuge, and for a shelter from storm and rain.* (Isaiah 3:4-6)
>
> *Because indeed, because they have seduced my people, saying "Peace!" When there is no peace—and one builds a wall, and they plaster it with untempered mortar—"say to those who plaster it with untempered mortar, that*

it will fall. There will be flooding rain, and you, O great hailstones shall fall; and a stormy wind shall tear it down. "Surely when the wall has fallen, will it not be said to you, 'where is the mortar with which you plastered it?'" Therefore, thus says the Lord God: I will cause a stormy wind to break forth in my fury; and there shall be a flooding rain in my anger and great hailstones in fury to consume it. "So, I will break down the wall you have plastered with untempered mortar. and bring it down to the ground, so that its foundation will be uncovered; it will fall, and you shall be consumed in the midst of it, then you shall know that I am the Lord. "Thus will I accomplish my wrath on the wall and on those who have plastered it with untampered mortar; and I will say to, 'the wall is no more, nor those who plastered it, that is, the prophets of Israel who prophesy concerning Jerusalem, and who see visions of peace for her when there is no peace;'" says the Lord God. (Ezekiel 13:10-16)

The harp and the strings, the tambourine and flute, and wine are in their feasts; but they do not regard the work of the Lord, nor consider the operation of His hands. Therefore, my people have gone into captivity. Because they have no knowledge; their honorable men are famished. And their multitude dried up with thirst. Therefore, Sheol has enlarged itself and opened its mouth beyond measure; their glory and multitude and their pomp, and he who is jubilant, shall descend into it. (Isaiah 5:12-14)

So, it appears that the ancient Hebrew prophets were very much concerned with atmosphere and its effects upon them. There are other scriptures that address this subject as well, and based upon the Bible's content, I can infer that they had various life cycles for different phenotypic outcomes and that building engineering that controlled environmental conditions and biological characteristics of human beings.

For some reason, this work was abandoned, and the scientific knowledge hidden in Bible codes throughout the Bible.

Finally, am I suggesting that we rid ourselves of modern conveniences and transportation? No! Rather that we redesign and enhance them to be compatible with our bodies. That people be engineered starting with the womb and then put on a life cycle. The time has come for a new psychology! Currently, we have science trying to understand nature without God and Christians trying to understand God without studying science. Current circumstances indicate to me that neither approach is working.

Chapter 8

METACOGNITIVE BEHAVIORISM

After science developed the scientific method as a tool for observing nature that guided the research enterprise, science became complacent. Admittedly, a tremendous amount of knowledge has been the result of the scientific method. However, there seems to be a crisis in processing this immense amount of information in ways that would improve our quality of life: The scientific method itself seems to have limits. Yes, our research methodology appears to have boundaries, some sort of ceiling in regards to processing known facts.

The microscope and telescope may have increased man's ability, to see, that is to observe phenomenon that is not observable without them but does nothing to increase accurate understanding or the big picture. Even though scientists in general may have intuitively understood this to be the case it would be the physicists that would feel the greatest impact. So, it was physicists who recommended that we think about how we process scientific information with Stephen Hawkins at the forefront. As a psychologist/general scientist working in the fields of psychology and human performance, I felt a tremendous pressure for a greater understanding of what I was working with. The fact that I was limited to making only direct observations in my work made it paramount that I have advanced methods of processing information available to me. Having a behavioral background, you might say that I have taken behavioral psychology to the next level.

I have increased the power of the observation process by thinking about how I think, actually taking control of how I observe natural processes and how I look at and organize known scientific facts that were also obtained by direct observations of human behavior. By improving my power to make direct observations, I am able to understand mind!

With an accurate understanding of mind, one is forced to see the hand of a creator who purposely designed and put in place a system for mind to come into existence and that this corresponds to the Biblical account of creation when properly understood. Adam the first behaviorist psychologist of the University of Paradise used direct observation to study the animal kingdom. Adam being the first creature to experience feelings having the need for a companion, proving that man before Adam was indeed autistic with the behavior pattern of what we now call animal. Man wasn't a social animal until Adam. The scriptures bear witness to my observation that mind is the autistic continuum.

The Two Witness – The Butterfly and Frog

The Bible teaches that more than one witness is necessary to establish suspicion, but only two or three are enough for an inquiry. The inquiry must be through and comprehensive. To this end, the creator has been true to this commandment as the butterfly and frog are the only two creatures available to us that prove that an organism can completely change form and function (behavior, diet, locomotion, and habitat). This also proves that the process can be accomplished in a short span of time. I am sure that Adam included these witnesses to evolution in his research with great profit. We can only ask and speculate if this rapid process of change exist in the reptile family only in regards to the animal species that includes humans.

I believe the answer is yes, and that is why only the frog is the only example so to prevent confusion. Also, the exaggerated change in mode of locomotion involved in the change to butterfly and frog, I understand as a teaching method for us so that we understand the importance of motion for human health and mind. I went to great lengths earlier to explain how human locomotion is so different from all other animals including primates and thus excluding us as primates along with other physiological factors. Human motion and posture as well as all the processes involved is mind!

Back to Eden

It appears that mankind has returned to the beginning, back to the days of Adam at the University of Paradise. Hopefully, Christianity will be restored back to its scientific roots as well. Yes, even though the Old Testament is full of scientific codes so is the New Testament. Remember that I stated earlier that Philo in his writings mentioned the early Christians that were known as Therapeutae for men and Therapeutrides for women. These early Christians devoted themselves entirely to the pursuit of knowledge and that their guiding principle for scriptural study was to interpret the Bible allegorically, they taught that the literal sense was symbolic of a reality hidden beneath the surface. I am a modern witness to this truth and a beneficiary of the method. We discussed earlier the Biblical principle of the two or three witnesses combined with a critical and thorough investigation. To this end, I would like to mention the work of George Sasson (1978) as another modern witness to this mode of scriptural interpretation. Sasson states that the Kabbalah was coded and used analogy, jokes, and puns as a means of hiding information. Myself and Sasson make two modern witness this satisfies Biblical requirements.

I am going to give one code from the New Testament to go along with the code I gave from the Old Testament to serve as two witnesses. This code deals with human performance and superhuman performance and shows how the codes also use variation as part of their format. According to the article, "Jesus cleanses the Temple – Commentary (2019), Jesus cleaning the Temple was a superhuman sign and that Jesus had to be more than a mere man. The cleansing of the Temple is written in John 2:13-15, and Nicodemus' visit and questioning Him about how He was able to perform this superhuman feat in John 3:1-15. Here are a few keys to the scientific discourse of this code dealing with the metamorphosis of Jesus:

> Enter the Kingdom of God = a physical form (phenotype) where special physical and mental performances are possible.
> Two Contrasting Environments
> 1 Water (02 bound to hydrogen)
> 2 Spirit – the Earth's atmosphere from the Earth's surface to outer space (air and oxygen)
> Jesus said the spirit (o2) going where it pleased was a play on words. He was distinguishing free o2 in the atmosphere from bound 02 in the molecule H2O. You hear the sound of it, but you do not know where it comes from or where it goes—a change in conscious states that facilitates a greater ability to use energy for Performances. Energy reserves released.
> Heavenly things – Advanced, hidden knowledge.

This knowledge was derived from the advance studies of Adam from his biological work while naming the entire animal kingdom God's Creation and was passed from generation to generation. Adam was the result of human genetic engineering. Adam was geographically isolated in order to change his phenotype and then his phenotype was passed on through his genes. Yes, Eden was an artificial environment designed to cause a certain species of creation to produce man, and man was isolated to become a special phenotype from environmental and dietary variation. We need to go back to Eden!

Chapter 9

IMPLICATIONS

The great theologian Matthew Henry in his commentary on the creation story Genesis 2:1-3, (1708, p.5) deduced that the creation of man was a gradual process, unlike modern theologians and preachers who believe man was the result of an instant act, immediately in the image of God. Even the New Testament informs us that we are lower than angels (Hebrews 2:6-7), so we aren't in God's image yet. The apostle Paul also makes it clear that man isn't currently in God's image (2 Corinthians 3:18). In the same chapter, Paul teaches that the Old Testament was veiled from the understanding of many Jews of Moses' times up to the time of Paul, and I can testify that much of the entire Bible is coded (veiled) today. Mathew Henry's insight is that there is a distinction between the words created and made, which was applied to the lower animals' creation and the word formed, which was said of man is important. From this Henry deduced that man's creation was a process.

In addition, one must consider other scriptures from the New Testament that help us understand this process. The first is Romans 4:17, it states:

> *(as it is written, "I have made you a father of many nations") in the presence of Him*
> *whom he believed——God, God, who gives life to the dead and calls those things which*
> *do not exist as though they did;*

The second New Testament scripture that bears light on this subject is Hebrews 2:8, and it reads:

> *You have put all things in subjection under his feet. For in that he put all in subjection under him. He left nothing that is not put under him. But now we do not yet see all things put under him.*

The book of 1 Corinthians the 15th chapter informs us that there is an order to this process and that we must first bear the image of Adam. Adam and his descendants lived almost a thousand years. We have got to get back to Eden and the tree of life before we can ever reach the image of God. The book of Revelations chapters 1-3 tells us that the church under the guidance of Jesus has access to both the hidden manna (knowledge) and the tree of life. This process is to take place on Earth! Let us read 1 Corinthians 15:25-26:

For He Must reign till He has put all enemies under His feet. The last enemy that will be destroyed is death.

The sciences during the age of Adam advanced to a very high level in fact, they possessed so much knowledge that they also were filled with so much pride that they could not acknowledge God. The Bible says as a result their biological nature was changed, and they were destroyed. They were allowed to advance in knowledge; however, they didn't advance in the spiritual man. Our Bible teaches that Jesus has paved the way so that we can now have the capacity to advance in character and knowledge.

I am hopeful that we as a people will began to change our perspective in regards to human life and human potential. It is my honest desire and prayer that my generation will be able to come together for the benefit of mankind. Hopefully, the ability to believe in a creator, the true Creator will stimulate and guide man into a new era where life will be a blessing to all on Earth. Both science and theology need to get on the right track. It should be apparent that planet Earth is a perfect laboratory for a very large variety of life and that man is a part of what we call the animal kingdom. The purpose of simple life forms is to produce more advance forms. The Bible with its sacrificial system allows animals to die for the sins of man. In principle, this is saying that the lower animal can be held responsible for man's nature, and this would imply that man came from the lower animal forms. In addition, in scriptures, when God punished man in general or specific nations the animals were also punished. In ancient Israel, there were laws regulating the relationship between man and animals, women and animals, as well as those for men and women. The Bible tells us that all the animals were created before man and that man was formed.

Like Mathew Henry's insights into the Word formed, I also get additional insights from the Word formed. Our bodies are precisely designed so that we have a different mode of locomotion that causes us to have a vestibular with properties that cause us to exhibit behavior that we call mind. The butterfly and frog teach us that, in nature, small, simple life forms can suddenly change into larger creatures with different forms and behavior. A process that through a specific species could have easily led to the sudden emergence of man. This is a process that neither the scientist nor the theologian can refute. It appears that even the ancient Egyptians were involved in human engineering as pictures remain of humans superimposed on numerical graphs, which indicates to me that they were studying human form and probably engineering human form for various purposes. We would be wise to control human form through growth cycles. Much research needs to be done in this area. My successful work in this area opens the door and proves that this would be fruitful adventure.

We can program the human life cycle for predictable outcomes for both physical and mental health with corresponding physical human performances. Planet Earth is the perfect laboratory for the process of the creation of man and mind as a human form with its capacity for obtaining and storying knowledge is the platform for a purposeful evolution and advancement of humankind aided by the advance knowledge of the Creator.

In order for us to succeed theology is going to have to realign itself with the teaching of the Bible. The Book of Ecclesiastes Chapter 3, verses 18-21 reads:

> *18 I said in my heart concerning the estate of the sons of Men, that God might manifest them, and they might see that they themselves are beast. 19 For that which befalleth the Sons of Men befalleth beasts; even one thing befalleth them; as the one dieth, so dieth the other; yea, they have all one breath; so that a man hath preeminence above a beast: for all is vanity. 20 All go unto one place; all are of the dust, and all turn to dust again. 21 Who knoweth the spirit of man that goeth upward to the earth?*

Like the preacher, it is my desire that it will be revealed to the theologian that man is an animal and to the scientist that there is a creator who designed our solar system so that it would produce mind at a level that we can't yet perceive. That the laws of multiple phenotypes apply to the human animal, just as it does to the rest of the animal kingdom, and this situation, if understood and applied to the human life cycle, will permit us to solve and overcome our pending national and global crisis.

Return to a Rational Faith

Currently, it appears that both the scientific and theological communities are bound and confined with self-imposed limitations, even though; not intentionally but nonetheless crippling. The goal of science is to create better living conditions for a limited mankind with the aid of the scientific method and that of the church is the return of the Creator until the Creator returns to save man on Earth and thus stand by idle while living conditions on Earth become hellish. We can see that the preacher or writer of Ecclesiastes was a scientific researcher in several disciplines. His major complaints were that man was limited by an arrested state of evolution where time was a restricting over ruling factor of life. Life was short, and death the ultimate reality. The preacher labored against the ignorance of his time and culture. The inability to see that the answer to their problem of short life span and death was to be found through the study of biology and recognizing the fact that man is a biological creature, just like the rest of God's creation and subject to the same laws presented to all life by the same laboratory called Earth.

Even the great preacher was limited in his time by the advanced knowledge being veiled by Moses, but today, that veil is beginning to be lifted. According to the writer of the book of Hebrews, the early church had lost the knowledge pertaining to its own theology.

The problems that threaten our world today and have reached crisis proportions are by no means in danger of being solved by the church or science in their current state. Lewis (1929) had this to say about the church:

> *The mighty works of the gospel accounts are made credible by being looked at in the light of Christ's own unique character. The apostolic mighty works may be made credible on the same basis. It is often objected to their authenticity, that they ceased with the apostles. That is true only as we greatly restrict the connotation of the term "Miracle". Anyone who knows how quickly the sub-apostolic church settled down into a well-regulated organization is not surprised at the limitations on its power but even then "the Miracles of grace" at least did not cease. And there is more than a suspicion among certain earnest Christians to-day that the loss of the power of the apostolic commission may yet be remedied. No one will question that it is the business of the church to encourage social and medical and psychological science to the utmost extent. But it is more than possible that its obligation goes even beyond this to the point of demonstrating a fresh that there are aspects to reality that cannot be reduced to science, and reserves of power that cannot be tapped by bare reason. It is such aspects to which religion answers, and it is such reserves of power, of which prayer and faith are the keys.* (p.930)

This describes a church without power, knowledge, and participation in the world's current crisis. Also, a church in denial if it expects science to come to truth or discover truth that is hidden in the Bible. As humans, we can only do what we can do when we can do it. How can an organization take two steps at once? How can the church recognize scientific truth and even reason beyond science when it isn't concerned with truth or reason beyond the first level of scriptures? We are left with a church that has strayed from its roots as a result it is without power or knowledge and a science with so much knowledge but without means of obtaining a synthesis or harmony with it. Lewis didn't explain the "grace" that remain with the church; however, in the power of prayer may a light shine bright enough so that there can be union of church and science. May the church return to its early form and may there be a re-emergence of the Therapeutae for men and Therapeutrides for women. Men and women in search of the Creator and His knowledge.

Book of Hebrews

Finally, I would like to end my discourse with a brief discussion of the deep level of the New Testament book: Hebrews. I won't discuss the entire book, but the parts that concern the functioning of mind. In this regards Hebrews talks about Mental functioning as related to the vestibular system. The author makes it clear that the Levitical priesthood didn't concern itself with bones, bone marrow, and mental states, but it did address muscle mass and blood health. In Hebrews 4:11-13, we have a code dealing with mind states that prevent us from living productively in the work of God:

The Word of God = most important principle
Hearing = vestibular system
Vestibular - first system to develop
Vestibular - controls immune system that develops in bone marrow
Immune system controls mind states; mind states influence immune function
Mind states (anxiety, depression) influence
Vestibular function and the Immune system
This is a double-edged sword.

This group of Hebrews had fallen short of the purpose of the church in that they couldn't function in the New Covenant, which concerned itself with mind states that allows people to prosper in the study of God's handiwork (the study of what we call science: knowledge combined with wisdom). These Christians met with disasters, and they couldn't understand the fundamental principles of Christianity. The church it seems has never recovered or corrected this initial blunder. However, the Creator has allowed knowledge to continue to increase, even though in an incomplete manner. It isn't too late; we can correct our course!

With advance knowledge existing for thousands of years, it would not be impossible for advance civilizations to have existed as well. Advance is also possible for us. There is no reason to contrive "aliens" for advanced civilizations of the past, nor do we need "aliens" for the present and future. We need a unity of knowledge and a unity of science and theology. Hopefully, I have been fruitful in providing a foundation for this to occur. With the knowledge that has been revealed in this book, it should be clear that we are the planned result of a creator who created the universe so that mind in a human form would evolve. The entire animal world was created before us so that a primitive vestibular system could evolve into a superhuman form thus proceeding through a biological process as the result of multiple phenotypes in the human kingdom. We were created to become superhumans, not angels. This biological process had to take place in the framework of time and environments with time as a 24-hour cycle, controlling the biological and chemical functioning and the learning and programmed behavior in sequential environments and body forms (species) until the human form evolved with behaviors that we call mind. So, biological forms had to proceed from primitive to advance. Basic laws of motion such as "relativity" and "uncertainty principle" in regards to the human forms allowing for different chemical expression and anticipation, which we call subjective reality. This gave the vestibular system the power to extend the autistic spectrum with what we call normal behavior as well as gifted and creating restrictions, limitations, and disabilities.

It appears that the ancient Greek philosophers might have been on the right track in regards to what Aristotle and Plato presented as "forms," even though their work has been passed down to us in ways that are unintelligible to the modern mind. Hopefully, if there is interest, I will be able to present the exact stages of the human life cycle as to more precise stages and "forms" and changes in forms and function of humans that facilitate health, mental health, and extreme longevity. It is important to realize now that a baby developing in its mother's womb can be negatively impacted by circular forces that will in fact program its nervous system to change the way that it perceives motion at a certain age period in life and thus cause schizophrenia, which is a motion disorder. There are other forces that can cause other developmental problems that can be experienced at different points in time of the nine-month period. A precise diet of certain proteins and vitamins can also determine the mind states of the child during this important nine-month window. It is within our power to bring almost every child into this world healthy.

Physics of Mind

Understanding how physics works in relation to mind states is fundamental to understanding mind and mental health. In this regard, I am presenting my last figure called "Physics of Mind." It is a brief outline of the relationship between biological forms (including changes in nervous system tissue), and forces that a human form is subjected to and which it is capable of producing. Einstein's relativity is itself relative and is inclusive of all forces and includes the uncertainty principle in regards to mind and remember that everything exists in relationship to mind in our entire solar system. Relativity exists in the form of a spectrum and function in regards to form and velocities, and without this, there is no mind. Both relativity and the uncertainty principle help to make mind possible as they are part of the functional principles of the vestibular and the rest of the nervous system. Without the uncertainty principle, the ability of mind to anticipate would not exist as the vestibular anticipates as a redundancy at higher and higher levels. Please note the following, it is an example of using mechanisms of schizophrenia as an element for enhanced understanding and creativity not just a symptom of disease. A component of this disease state known as "alteration of concept formation" gives insight into the workings of the vestibular system ability to manipulate information (data) in the form of mind templates, demonstrating that schizophrenia and creative problem solving are the same processes with different outcomes. It is a snapshot of psychological anticipation and distortion, as I am manipulating ideas as part of a template to make clear how principles of physics operate psychologically. However, since this is a snapshot, I will have to continue this topic in a volume format with the theory and examples of the mind processes becoming clearer as I proceed. You can actually see how my mind has manipulated data continuously from the start of this book to end. The following is an example of a specific template in which the forces of both classical physics and quanta mechanics are stated in ways and a format that are both novelties as examples of the working of the mind.

Physics of Mind
Mind – Motion, Matter, and Mass

Paul (Acts 17:28)
For in Him we live and move and have our being
Law of distortion (Perceptual) – The three laws of the harmony of Motion

1. Relativity – Large forces, Masses, and Motion
2. Uncertainty Principle – Small forces, Masses and Motion
3. Mind (vestibular) – Intermediate forces.

Masses, and Motion (biological)

Motion = Life
Motion (redundant) = Mind
Motion-Ruled by geometry (Math)
Geometry-Basic Mind

Distortion

Distortion – (asymmetry within the nervous system) – Biological Relativity
Uncertainty Principle – Subjective mind
Biological asymmetry – Perceptual distortions
(Change in electrical/biochemical properties of Nervous System)
Vestibular = Function within the range of Micro accelerations and Micro psychological stress

Macro Acceleration and Mind

- 40 Miles per hour – stress for vestibular
- Accelerations by Motion in Curves or Angles – stress for vestibular
- Macro Psychological Stress – Overwhelms vestibular

Relativity Spectrum – each spectrum is a distorted form of energy in regard to all other forces and forms. The redundant motion spectrum:

1. Movement of the solar system
2. Movement of light
   - a electricity
   - b magnetism
3. Movement of the earth's atmosphere
   - c change in pressure
   - d change in weather
4. Movement of the earth – gravity
5. Movement of the animals (vestibular controlled)
   - e biological relativity: asymmetry in the nervous system
   - f biological uncertainty principle: perceptual properties based on anticipation of motion
   - g mind in human form – biological speed
6. Relativity – large speeds
7. Uncertainty principle (quantum mechanics)
8. The unity of forces is that all forces are distorted forms in regard to all other forms of energy.
9. Without this spectrum, there could be no reality as we experience it and no mind.

Now that we know what we are and why we exist it, becomes our duty to become what we are supposed to be. We have the potential to do just that. Science even though independent of the church has nonetheless developed enough knowledge when properly put together and understood. The church has been a custodian for the scriptures and in modern times a support to society with prayer and good works. As for me, I have remained an independent researcher profiting from the vast knowledge of science and assistance of scriptures meanwhile avoiding hindrance from science and the church. The time has come when all can benefit from an advance in the biological sciences of epigenesis and epigenetics without a recourse to the understanding of how biology is affected by quantum physics as suggested by a leading biologist (Lipton, 2005). The health goals that we are seeking are not to be gained by subscribing to eastern medical science, but from a true understanding of the science of energy, mind and forms left to us from the ancient west. The proper understanding of growth rate, form, and function is utmost importance and the quantum physics will take care of itself. With the health benefits and growth in intellect over time, we may be able to understand mind at the quantum level, even though, there is no need to do so.

It appears that motion not under the control of the vestibular system; let us call it artificial motion (technology) has a negative effect on the vestibular similar to that of microgravity (altered sensory stimulus state). Motion and human biology in regards to human health is a subject not studied from every perspective.

Hopefully, this won't remain the case, and that research will give us a complete and accurate understanding. Meanwhile, I will continue to personally pursue this issue. Are accelerations above 40 miles per hour not only unmanageable for schizophrenics but a serious problem for all humans?

The economic demands created pressure on the western world hundreds of years ago, and this led to a rush for artificial motion as a mean or aid to the economy, and this was without a working knowledge of the true nature of man. And this correlated with the increase of mental illness in Europe at the time. Unfortunately, one of every four children in the US today suffers from mental illness. We need to be careful how we use modem motion technology, and transportation needs to be made safe for human beings!

Final Note

I watched a "science channel" program once about the history of the US space program. When confronted with the problems of humans in space several prominent scientists suggested that astronauts could be genetically engineered for the purpose. I have no knowledge of how NASA approached this idea, or even if they did, however, based on my own work, I am sure that this is possible. Unfortunately, we are engineering mental illness in the US. It isn't an experiment, but a human disaster caused by a lack of knowledge. Epigenetic and epigenesis is real, and it applies to the human condition not just animals from the lower species.

At least 14 million Americans are expected to experience mental illness this year, and this number really doesn't project the millions of black Americans who never encounter the mental health system but are highly impacted by what (Noonan, Mondragon & Wagner, 2016), call epigenetics factors. At the same time, it is through epigenetics or human engineering that there is hope and help for all.

You have probably noted that I have been faithful to figure one in covering the subject of mind from physics to theology. Theology proves to be very important in that both Psalms and Hebrews considers the vestibular as the primary principle in life. I have done my best in this book to reestablish this fact. It was the prophet Ezekiel 47:1-9, who connects movement with life and codes for a movement program where human movement changes from terrestrial bipedal, to aquatic quadrupedal from vertical to horizontal. In other words, a movement program for humans necessary for life! That at certain stages of life, movement must make dramatic change for life to persist.

The Plan

The current worsening of the mental health status of Americans is not only a human disaster but an actual threat to our national security. With the current mental health crisis in the US and the inability of our mental health, educational, and criminal justice systems to help, they are helpless and in denial.

In addition, our churches are in retreat from all social problems, looking forward to the Rapture. Making bad matters worse, our greatest scientists believe that evolution is leading to disaster. This book was written to inform, empower, and serve as a source of good news. Even though our current constitution is a document concerned with adults, as it is believed that birth is a process treat all the same with equal rights in regard to the pursuit of happiness and prosperity, nothing could be farther from the truth. Life developing in the womb is a very dangerous enterprise, and its proper development is critical for life after birth. This is just the first step for a good life after birth, but the most important one. So, the most important thing I am hopeful for is a change in our constitution, giving rights to the unborn life while in the womb.

Currently, different culture groups have children for different reasons:

1. Increase family numbers for economic survival
2. For a welfare check and food stamps
3. Make a boyfriend or husband happy
4. Proof of womanhood or manhood
5. Unintended pregnancies – no planning

We now have a large economy, though I would argue not large enough, but we have a large population and a very large portion of the population is mentally impaired with that portion on the increase. With this increase in population and economy, it is imperative that we shift attention to child bearing and make sure that we have the proper reasons, medical care, and resources devoted to this life process.

With so much at stake and so much to be done and the necessity to change both public and government opinions, the constitution, and the way of life of an entire nation, my efforts are in need of much prayer. Prayer, writing and to publish for now. One step at a time.

Chapter 10

THE NATURE OF MIND (SUMMARY)

The human mind fundamentally is based on a moving system that makes observations. The observer is the vestibular system. In order for the vestibular to make observations, it must be stable or unmovable, and it also produces motion and mind from the stable position produced by l-G gravitational field. With a proper diet and metabolism, the proper chemicals are available in proper amounts for the vestibular to control and operate the nervous system with electrical signals. The vestibular control all mental and biological processes by observing the movement of light (the 24-hour variation in blue light intensity). The variation in posture brought about by evolutionary processes such as metamorphosis such as motion in a horizontal posture to that of motion in a vertical posture. The computational properties of the vestibular system at each stage of motion and posture as demonstrated in the seizure called the Jacksonian March correspond with the different states of mind from the beginning mind of the autistic without a will, to a fully functioning subconscious and subjective conscious willful mind. It is this difference in posture and gait that separates us from the rest of the animal kingdom. A certain level of mind corresponds with a certain posture and gait. Man, when first changing form and posture from a primitive reptile to a primitive man, was completely autistic, fearful, without a will, driven by selfish survival instincts carried out by savant skills: map, calendar and math calculating.

As reflexes, posture, and gait improved and man grew in size a will and subjective thought also emerged. Now mind contains all of these subsets of behavioral templates stored away in the human nervous system. As a result, we have what I termed mind as the autistic spectrum, and all these spectrum templates are under the workings of either a functional or damaged vestibular system. The capacity of the vestibular system is extremely vulnerable while the child is developing in the womb. The inability of the vestibular system to handle motion after birth can be programmed before birth, causing the vestibular electrical stimulation to other important nerve tissue to be abnormal causing distortions in their shape and function when dealing with motion at the fundamental functions, and this can cause the more primitive behavior templates to become functional at the expense of the more advance ones. This is how we get developmental, behavioral, and mental disorders. Of course, we can and should control this process. Even after birth, the vestibular system can be harmed by the forces in our environment, both natural and manmade.

Our minds are limited in more than one way. First, mind goes through its full evolutionary states every 24 hours as it repeats it horizontal stages via sleep and then the vestibular changes the posture to vertical with its will and subjective thought process intact. The mind that we are so proud of last for only a fraction of the 24-hour repeating cycle.

Thankfully, these properties of our body/mind, makes it possible for us to have mental health for all of us. Of course, as modified reptiles with our unique bipedal locomotion in a vertical posture requires a tremendous amount of vestibular computing power for motion and at this amount of nervous system control gives us an additional power that we call mind. This mind's fundamental concern is the safety and survival of both mind and body.

So, a healthy mind also requires a safe environment. Any problems, and the vestibular can only activate primitive behavior templates, which we know as mental illness. In order to maintain a healthy mind, mind must perceive itself as safe. Is this all an accident?

Chapter 11

ANSWERED PRAYER

Several thousand years ago a scholar, King, and Superman named David wrote Psalms 19:

Psalms 19
THE heavens declare the glory of God:
And the firmament shows His Handiworks.
2. Day unto day utters speech,
And night unto night reveals Knowledge
3. There is no speech nor language
Where their voice is not heard.
4. Their line has gone out through all the earth,
And their words to the end of the world.
In them He has set a tabernacle for the sun,
5. Which is like a bridegroom
coming out of his chamber,
And rejoices like a strong man to run its race.
6. Its rising is from one end of heaven.
And its circuit to the other end;
And there is nothing hidden
from its heat.
7. The law of the Lord is perfect,
Converting the soul;
The testimony of the Lord is sure,
making wise the simple;
8. The statues of the Lord are right,
rejoicing the heart;
The commandment of the Lord is pure,
enlightening the eyes;
9. The fear of the Lord is clean,
enduring forever,
The judgements of the Lord are true and
righteous altogether.
10. More to be desired are they than gold,
Yea, than much fine gold;
Sweeter also than honey and
the honeycomb.

11. Moreover by them Your
Servant is warned.
And in keeping them there is
great reward.
12. Who can understand his errors?
Cleanse me from secret faults.
13. Keep back your servant also
from presumptuous sins;
Let them not have dominion
over me.
Then I shall be blameless,
And I shall be innocent of
great transgression.
14. Let the words of my mouth and the
meditation of my heart
Be acceptable in Your sight, O LORD, my strength
and my Redeemer.

It was during the time of the prophet Isaiah that the Old Testament church dropped the scientific enterprise (God's handiworks). However, a small group kept science alive, even though separate and independent from the church in general. Besides misery, the very people who wrote the Bible lost the keys for Bible study and understanding leading them and the world into darkness. Unfortunately, this state of affairs has continued to this day.

Thankfully, due to the eternal goodness of the Creator, light has begun to shine. Not only is this book an answer to David's prayer, it is a modern English interpretation of Psalms 19:

It is due to the precise nature of the solar system and terrestrial laws of gravity and motion that biological life and mind exist. All the functions of the mind and body are under the control of 1-G gravity and the movement of the sun and blue light spectrum. Body and mind states are subject to a 24-hour repeating cycle. The mind proceeds through sequential stages, step by step as different mind states in 24-hour repeating cycles, and this biological law facilitates mind. Even though the mind is a wonderful creation, it is fragile; reveal to me its workings.

We know that this prayer was answered as David was employed by King Saul to treat his depression. Unfortunately, our modern world is in a severe mental health crisis as it lacks this knowledge. Due to an insufficient understanding of evolution and the scripture, both science and Christianity are at odds and helpless when it comes to understanding mind and can't offer help. The law in the scriptures requires knowledge of and compliance to the laws of God's handiworks before they can be understood and obeyed.

Understanding God's handiworks is fundamental and foundational before we can have success with any educational system, including Christian education. I have been praying, studying and doing research since the age of four in search of truth. My prayer was answered. This truth, I have been trying to share. If divine patterns or principles are known and followed in all of life activities then wonders and miracles follow. There is a divine order for all things. There is hope for America and mankind!

BIBLIOGRAPHY

1. Atwater, P.M. H. (2013). Future Memory. Charlottesville: Hampton Roads Publishing Co., Inc.

2. Awad, A.G. & Ritsner, M.S. (Eds.) (2007). Quality of life: Impairment in schizophrenia, Mood and Anxiety Disorder. Netherlands: Springer.

3. Barnes, C. (1988). Melanin: The chemical key to Black greatness:
The harmful effect of toxic drugs on Melanin Centers within the Black human. Houston Texas: C.B. Publishers.

4. Barnes-Svamey, P. (Eds.). (1995). The New York Public Library: Science Desk Reference. New York, New York: Simon & Shuster Macmillan Company.

5. Bergman, P.G. (1992). The riddle of gravitation. Toronto, Ontario:
General Publishing Company.

6. Berne, R.B. & Levy, M.N. (Eds.; 1990), Principles of Psychology. Baltimore: The C.V. Mosby Company.

7. Bradshaw, J.L., Georgiou, N., Phillips, J.G., Iansek, R., Chiu, E., et al (1988).
Motor sequencing problems in Parkinson's disease, hurting's disease and Tourette syndrome 1: A review of basal ganglia involvement. In J.P. Piek (Ed.). Motor behavior and human skill: A multidisciplinary approach (pp. 305-315). Champaign, IL: Human Kinetics.

8. Brain Breakthrough: The vestibular system and brain (2014).
Retrieved August 1, 2015 from http://www.brainbreakthrough.com/vestibular.html

9. Brand, T. Schautzer, F., Hamilton, D.A., Bruning, R.,
Markowitsch, H.J. and Kalla, R. et al. (2005). Vestibular loss causes Hippocampal atrophy and impaired spatial memory in hnnisins Brain; 128, 2732-2741.

10. Campbell, R.J. (1981). Psychiatric dictionary (Fifth ed.). New York: Oxford University Press.

11. Carlson, N.R. (1988). Foundations of Physiological psychology.
Needham Heights, MA: Simon & Shuster, Inc.

12. Commentary – The fourfold Gospel (2019). Jesus Cleanses the Temple. Bible Study Tools, 1-3. Retrieved, May 22, 2019, from https://www.biblestudytools.com/Commentaries/the-fourfold-gospel

13. Dake, F.J. (1999). Dakes Annoted Reference Bible: The Old and New Testaments. Lawrenceville, Georgia: Dake Publishing, Inc.

14. Fischback, G.D. (1993). Mind and Brain. In reading from Scientific America: Mind and Brain (P.1). New York, New York: W. H. Freeman and Company.

15. Gamow, G. (1998). The great physicists from Galileo to Einstein. New York, New York: Dover Publications, Inc.

16. Gilbert, S. F. (2002). The Genome in Its Ecological Context: Philosophical Perspectives on Interspecies Epigenesis. In Speybroeck, L.V. Vande-Vizver

G., & DeWacle, D. (Eds.). From Epigenesis to Epigenetics:The Genome in Context. Annals of the New York Academy of Sciences:Vol 981 (pp. 202-218).

New York Academy of Sciences.

17. Goldstein, M. J., Baker, B.L., & Jamison, K.R. (1980).

Abnormal Psychology: Experiences, Origins, and Interventions. Boston MA: Little Brown and Company.

18. Hallahan, D., P., Kauffman, J.M. & Lloyd, J.W. (1999).

Introduction to learning disabilities. Needham Heights, MA: Allyn & Bacon.

19. Halstead, L.B. (1989). Brain. In R. Steel & A. Harvey (Eds.). The encyclopedia of pre-historic life (pp. 22-23).

20. Harari, Y.N. (2015). Sapiens: A Brief History of Humankind. New York, NY: Harper Collins Publishers.

21. Harris, S.S. (2006). Symposium. Optimizing Vitamin D intake for populations with Special needs: Barriers to effective food fortification and supplementation. Journal of Nutrition, 136: 1126-1129

22. Hawkins, S., & Mlodinow, L. (2010). The Grand Design. New York: The Random House Group, a division of Random House, Inc.

23. Henry, M. (1708). Matthew Henry's Commentary in one volume. In L.F. Church & F.R. Hist (Eds.). Genesis 11-3 (p.5). Zondervan Publishing House: Grand Rapids, Michigan.

24. Holarchy Coherent and Fractal – Bliss: Solutions to Coherence Wars: (2016 March 24). Harmonic inclusiveness/fractality. Predicts viability - in the Heart (HRV) and in every living oscillator. Retrieved February 7, 2017 from HRV-app.com

25. Jackson, J. G. (1993.) Man, God and Civilization. New York, New York: Carol Publishing Group.

26. James, G. G. M. (1989). Stolen Legacy. Newport News, Va. United Brothers Communications Systems.

27. Jones, S. (1993). The language of genes. New York, New York: Doubleday.

28. Kagan, G. (1999). Born to be shy. In R. Conlon (Eds.), States of Mind (pp.29-51).

New York, New York: John Wiley & Sons, Inc.

29. Kalat, J.W. (2007). Biological psychology. Belmont, CA: Thomas higher education.

30. Kasumyan, A.O. (2004). The Vestibular System and Sense of Equilibrium in Fish. Journal of Ichtyology, vol. 44, pp. 5224-5268.

31. King, it (1994). African Origin of biological psychiatry. Hampton, Virginia: U. B. & U.S. Communication Systems, Inc.

32. Keller, E. F. (2002). Developmental Robustness. In Speybroeck, L.V.,

Vande Vizren G. & DeWacle, D. (Eds.). From Epigenesis to Epigenetics: The Genome in Context. Annals of the New York Academy of Sciences:Vol 981 (pp.189-201). New York Academy of Sciences.

33. Lehninger, A.L. (1979). Biochemistry. New York, New York: Worth Publishers, Inc.
34. Lewis, E. (1929). The Abingdon Bible Commentary. In F.C. Eiselen, E. Lewis & D.G. Downey (Eds.). The miracles of the New Testament (pp. 921-930). Doubleday & Company, Inc.: Garden City, New York.
35. Lipton, B. (2005). The Biology of Belief: Unleashing the power of Consciousness, Matter, & Miracles. Santa. Rosa, VA: Mountain of Love/Elite Books.
36. Maier, P.L. (1999). Eusebius – The Church History: A New Translation with Commentary. Grand Rapids, Michigan: Kregel Publications.
37. Moyer, L. (2014, May). What is the Vestibular System and why is it important? Retrieved August 3, 2015 from SproutsDevelopment.com
38. Meyers, D. G. & Jeeves, MA. (1987). Psychology through the eyes of faith. New York, New York Harper & Row, Publishers, Inc.
39. Parker, O. (2012). Vestibular System Development in Babies. Retrieved August 1, 2015 from BebezClub.Com
40. Restak, R.M. (1994). The modular brain. New York, New York: Macmillan Publishing Company.
41. Roberts, J. E. (2015). Circadian rhythm and human health. Photobiology. 1-30.
42. Sasson, G. (1978). The Manna Machine. Facebook, (pp. 1-7). Retrieved May 18, 2019, from https://m.facebook.com/notes/the-voynich-manuscript-revisited/the...
43. Scripture taken from the New King James Version. Copyright© 1982 by Thomas Nelson, Inc. Used by permission. All rights reserved.
44. Shapiro, LA. (2002). Genome organization and reorganization in evolution: Formatting for computation and function. In Speybroeck, L.V., Vande Vizver, G. & DeWacle, D. (Eds.). From Epigenesis to Epigenetics: The Genome in Context. Annals of New York Academy of Sciences: vol. 981 (pp. 111-134). New York Academy of Sciences.
45. Smith, P.F. & Darlington, C.L. (2013). Personality change in patients with Vestibular dysfunction. Prant Hum Neurosci. 7:678
46. Spoor, F., Garland, T. & Walker, A. (2007). The primate semicircular Canal system and locomotion. Proc Natl Acad Sci USA 104 (26) 10808-10812.
47. Tills, D. (1989). Man. In R. Steel & A. Harvey (Eds.). Encyclopedia of Prehistoric life (pp.121-123). New York, New York: Gramercy Publishing Company.
48. Vestibular Dysfunction (2014). Retrieved August 1, 2014, from http://www.child-behavior-guide.com/vestibular-dysfunction.html.
49. Wagner, F.A., Velasco-Mondragon, H. E., & Noonan, A.S. (2016). Improving the health of African Americans in the USA: an overdue Opportunity for social justice. Public Health Reviews: 37, Article number 12.

50. Weiner –Vahern, S.R., Hamilton, DA, & Weiner, S.I. (2013). vestibular activity and cognitive development in children: Perspectives, frontiers in Integrative Neuroscience, 7, 92.
51. Wikipedia, the free-encyclopedia (2014, November, 7). Vestibular System Retrieved August 1, 2015, from http://en.wikipedia.org/wiki/vestibular-system.
52. Wilson, E. O., (1999). The diversity of life. New York: WW. Norton & Company
53. Winter, D.A. (1995, December). Human balance and posture control during standing and walking. Gait & Posture vol. 3, Issue 4. Pages 193-214. Abstract retrieved December 11, 2016 from https/pdfs.semanticscholar.org

CPSIA information can be obtained
at www.ICGtesting.com
Printed in the USA
LVHW081523100222
710600LV00011B/396

9 781637 643204